Neotropical Montane Forests
Biodiversity and Conservation

Abstracts from a Symposium at The New York Botanical Garden,
June 21–26, 1993

edited by Henrik Balslev

1993

T0163278

AAU REPORTS 31

Department of Systematic Botany, Aarhus University
in collaboration with
The New York Botanical Garden

INTRODUCTION

Neotropical lowland rain forests have received tremendous attention in the last decade, particularly in the debate concerning biological diversity. The lowland forests have been mentioned over and over again as examples of the genetic richness of the tropical forests, and it *is* true that in small sample plots of, for instance, one hectare of lowland rain forest, the number of species is enormous. Lately, however, it has become increasingly clear, that the high alpha-diversity of lowland tropical rain forests is not accompanied by an equally high beta- and gamma diversity.

Montane forests, on the other hand, have been known for their richness, but when estimating their alpha-diversity by traditional methods using quantitative inventories of trees, their species richness does not equal that of lowland forests. Recent work that includes other life forms than trees, *i.e.* epiphytes, climbers, shrubs, non-vascular plants, *etc.*, however indicate that the alpha-diversity in many montane forests equals that of lowland forests. When studying the diversity patterns on slightly larger geographic scales, it has become evident that differences in species composition between localities in montane forests is tremendous, *i.e.*, their beta- and gamma-diversities are high.

It appears that, although they cover much smaller extensions than lowland rain forests, montane forests are the homes of the better part of the biological diversity of the Neotropical countries.

Discussions and attention relating to conservation has been equally skewed towards the lowland forests. Still, the montane forests are just as threatened by road construction, agriculture, cattle farming, *etc.*, as the lowland forests are. Considering this lack of attention to montane forests, despite their importance for the biological diversity of the Neotropics, it was appropriate to call for a symposium addressing these issues. Much knowledge obviously exists, but often in a scattered form, and it was the intention that a symposium could bring together researchers and others with interest in neotropical montane forests biodiversity and conservation, to exchange ideas and knowledge, and to call attention to the importance of these ecosystems for the future of genetic resources of our planet.

The response to the first announcement of the **Neotropical Montane Forest Symposium** was enthusiastic and over 80 scientists indicated that they were interested in contributing a paper to the symposium; still more indicated their interest in participating.

The purpose of the symposium is to document biological and ecological diversity in neotropical wet and moist montane forests and place it in the context with other neotropical areas. Emphasis will be given to plants, with regards to taxonomic and ecological richness, threatened and conservation areas of high biodiversity and endemism, the evolution, diversity, speciation, distribution, and biogeography of Andean-centered genera and families.

SCIENTIFIC PROGRAM

The topics to be treated at the symposium include the following:

Vegetation and Flora
• General overview of biogeography and conservation
• Research studies (Costa Rica, Venezuela, Colombia, Ecuador, Peru, Bolivia, Argentina)

Palaeobotany
• Palaeobiogeography and palaeodiversity/vegetation history

Taxonomy
• Non-vascular plants, including fungi, lichens, hepatics, and mosses
• Vascular plants, including ferns, fern allies, and gymnosperms. Discussions of flowering plant families

Man-Plant-Animal Interaction
• Bird distribution patterns
• Plants and people in the montane region
• Human influence and plant uses

Altitudinal Zonation and Methodology
• Current problems and methods in analysing montane vegetation

Conservation and Protection
• Current problems and future needs in the conservation and management of montane forests

PROGRAM SCHEDULE

The overall schedule for the symposium, as it looks when this abstract booklet goes to press, is given below. The detailed schedule of the talks is found separately in the handouts.

Saturday, June 19, 1993
Arrivals and pickups to The New York Botanical Garden and transfers to Fordham University dorms
12.00–6   Registration in NYBG Auditorium

Sunday, June 20, 1993
Arrivals and pickups to The New York Botanical Garden and transfers to Fordham University dorms
9.00–12   Registration in NYBG Auditorium
2.00–6   Registration in NYBG Auditorium
1.00–1.30   Herbarium and Library orientation

Monday, June 21, 1993
9.00   Welcome (Long, Boom, Forero) – Auditorium
9.20–12   General session
12–2   Box lunch (NYBG) – picnic grounds, Twin Lakes
2–5   General session
6–7.30   Dinner – Fordham University
8–9   Keynote lecture: Kent Redford – Auditorium
             "Conservation of Biological Diversity in a World of Use"
9–10   Social hour – Rotunda

Tuesday, June 22, 1993
9–12   General session, mostly ecological – Auditorium
12–2   Box lunch – NYBG
2–5   General session, mostly ecological – Auditorium
7.00   Banquet – Snuff Mill

Wednesday, June 23, 1993
9-12   General session, mostly phanerogams – Auditorium
12–2   Box lunch – NYBG
2–5   General session, mostly cryptogams – Auditorium
6–7.30   Dinner – Fordham University
8 – 10   Special evening session – Auditorium
             "Coca, Cocaine, Opium Deforestation and its Conservation
             Implications"

Thursday, June 24, 1993
9–12   General session, mostly phanerogams – Auditorium
12–2   Box lunch – NYBG
2–5   General session, mostly ecological (altitudinal zonation)
6–8   Dinner – Fordham University
8–   Discussion based on altitudinal zonation talks – Auditorium

Friday, June 25, 1993
9–12   General session, mostly ethnobotanical and conservation –
          Auditorium
12–1.30   Box lunch – NYBG
1.30–5   General session, conservation and protection
4–4.30   Summation
4.30–5   Final words from Committee and salutations

Saturday and Sunday, June 26–27, 1993
9–5   Library and Herbarium available for study

ORGANISATION

The Neotropical Montane Forest Symposium is sponsored by:

Institute of Systematic Botany, The New York Botanical Garden,
Bronx, NY 10458-5126, USA.
Telephone 1-718-817-8645/8654, Fax 1-718-562-6780

and

Department of Systematic Botany, Institute of Biological Sciences,
Aarhus University,
Nordlandsvej 68, DK-8240 Risskov, Denmark.
Telephone 45-86-210677, Fax 45-86-211891.

The leaders of these institutions, Director of the Institute of Biological Sciences, Aarhus, **Ivan Nielsen,** President of The New York Botanical Garden, **Gregory Long,** Vice President for Botanical Science, NYBG, **Brian Boom,** and Director of the Institute of Systematic Botany, NYBG, **Enrique Forero,** stood behind the organising committee with moral, economic, and organisational support.

The organising committee included **James L. Luteyn** (chair, NYBG), **Steven P. Churchill** (NYBG), and **Henrik Balslev** (Aarhus). **Elvira Cotton** was assistant to the organising committee.

ACKNOWLEDGEMENTS

The organising committee and the sponsoring institutions wish to acknowledge the support received from the following sources:

National Science Foundation (USA)

The John D. and Catherine T. MacArthur Foundation (USA)

The Andrew W. Mellon Foundation (USA)

Danida, (Danish International Development Assistance), Copenhagen

Natural Science Research Council (Denmark)

The Nature Conservancy (USA)

United States National Herbarium, Washington

Field Museum of Natural History, Chicago

Missouri Botanical Garden, St. Louis

Hugo de Vries Laboratory, Amsterdam

World Travel Specialists, Inc., New York

ABSTRACTS

The following pages include the abstracts of the presentations given at the symposium, as lectures or as posters.

The abstracts are arranged in alphabetical order by surname of the first author.

The abstracts have been edited for clarity and for consistency with the format of the abstract booklet. Since there has not been time to pass the edited versions by the authors for approval, any error accidentally introduced in the editing process is the responsibility of the editor.

LECTURE

Tropical and Sub-Andean Epiphyllous Bryoflora in El Macizo de Tatama, Colombia

J. Aguirre-C. and E. Castillo L.

We studied the epiphyllous bryoflora in the central part of the Colombian Cordillera Occidental on the western slope of Macizo de Tatama (550–1650 m; 4°50'–5°15'N, 76°12–27'W) and found 111 species of bryophytes divided between 98 hepatics and 13 mosses. The dominant family was Lejeuneaceae with 72 species and 30 genera. The highest species richness and coverage values were found at 1090 m and 1350 m elevation. The leaves on which the bryophytes were found fell in the mesophyllous and macrophyllous size classes and their surfaces were glabrous. The best represented species were: *Odontolejeunia lunulata* and *Cyclolejeunea angulistipa* (Lejeuneaceae). Other well represented species with high coverage values were: *Odontolejeunea convexistipa*, *Drepanolejeunea inchoata*, *Drepanolejeunea crucianella*, *Leptolejeunea elliptica* and *Ceratolejeunea maritima* (Lejeuneaceae) and *Crossomitrium patrisiae* (Musci) and *Metzgeria* sp. (Metgeriaceae). The average number of bryophyte species growing on a leaf was 13 and the highest number of species found was 31.

Jaime Aguirre-C. and Enrique Castillo L., Instituto de Ciencias Naturales, Universidad Nacional de Colombia, Apartado Aéreo 7495, Bogotá, Colombia.

POSTER

The Vegetation and its Uses in the District of Pamparomas, Peru

J. Albán

The purpose of this study was to investigate the ethnobotany of the rural community Pamparomas on the western slopes of Cordillera Negra in the northern part of the Department of Ancash, located at 1500–5000 m elevation. The study included both wild and cultivated plant species and the information collected concerned their vernacular synonymy, their habitats, and vegetation types where they grow, taking altitudinal and climatic zones into account. Of the 22 villages in the district 16 were visited during this study. Specimens were collected for the correct botanical identification. The vegetation types were studied with 50 m transects and 10 × 10 m plots. In total, 1200 specimens were collected. Until now 150 species in 46 angiosperm families and 15 cryptogamic species have been identified. The best represented families are Asteraceae, Solanaceae, Lamiaceae, and Rosaceae. The species are being grouped according to their uses in: edible plants, spices, medicinal plants, fodder plants, artesanal plants, and plants for construction.

Joaquina Albán, Departamento de Etnobiología, Museo de Historia Natural, Peru.

POSTER

Altitude Zonation of Volcanic Mountain Vegetation on Popocatepetl, Mexico

L. Almeida-Leñero, A. Herrera and A. M. Cleef

The purpose of this work is, within the UIBS-UNESCO *Comparative Mountain Ecosystems Study*, to understand the functions of tropical mountain ecosystems on Popocatepetl and compare them with other tropical mountains. Popocatepetl was chosen because of its marked altitudinal gradient, its climatic and topographic diversity, and its biogeographical importance (it belongs to the transition zone between the Neartic and Neotropical regions). These factors determine the existence of different plant communities. The methodology used was that of the Zürich-Montpellier school, adapted for the tropandine mountains of America. It is based on altitudinal transects, chosen with cartographic analysis looking for the less disturbed areas. Samples are taken every 100–200 m, distinguishing between zonal and azonal vegetation. The average sample size is 500 m². In each stand physical data are estimated: slope, orientation, topography, exposition, and type of soil. A complete vegetation inventory is obtained, the cover is expressed in percentage, this information is then tabulated for interpretation with environmental values. This method characterizes and defines important biogeographical areas, determines floristic composition, and gives a better understanding of the relationship and dynamics existing between these communities. The resulting data may be used for planning phytogeographical and ecological research. In this interval (2840–3830 m), 24 relevés were carried out: five in the *Abies religiosa-Cupressus benthami* association (2840–3050 m); seven in the *Abies religiosa* association (3070–3340 m), and 12 in the *Festuca tolucensis-Pinus hartwegii* association (3460–3830 m) with two sub associations *Geranium potentillaefolium* (3460–3620 m) and *Cirsium jorullensis* (3770–3830 m). In the first community, 35 vascular species and five non-vascular species were found in 27 and nine genera respectively. In the second, 38 vascular and 17 non-vascular species in 30 and 15 genera respectively. In the third community, 35 vascular and 21 non-vascular species in 22 and 16 genera. The best represented plant families in these forest are Asteraceae, Poaceae, and Caryophyllaceae. Comparing the three associations, *Abies* forest has the highest diversity because its soils contain most organic material and has the highest edaphic and climatic humidity.

L. Almeida-Leñero and A. Herrera, Laboratorio de Biogeografía. Facultad de Ciencias, Universidad Nacional Autónoma de México. c.p. 04510, Mexico. Telephone 6-224836, Fax 6-224828. A. M. Cleef, Hugo de Vries Laboratory, University of Amsterdam, Kruislaan 318, 1098 SM Amsterdam, The Netherlands. Fax 31-20-5257715.

LECTURE

Diversity and Origins of Andean Rubiaceae

L. Andersson

The paper is based on data from an herbarium and literature inventory of neotropical Rubiaceae. The purpose is to describe the diversity of Andean Rubiaceae and to trace their evolutionary origins. A total of 75 genera and 645 of the Rubiaceae have been recorded from the tropical Andes above 1,000 m altitude. The largest genera in the Andes are *Palicourea* (164 spp.), *Psychotria* (129), *Manettia* (71), *Galium* (30), *Rudgea*, *Faramea* and *Hoffmannia* (21 each), *Ladenbergia* (18), *Arachnothryx* (16), *Borreria* and *Arcytophyllum* (14 each), *Gonzalagunia* (13), and *Cinchona* (12). Generic endemism is low in the Andes, only the three small genera *Dioicodendron*, *Pimentelia*, and *Wernhamia* being strictly endemic and another five (*Arcytophyllum*, *Cinchona*, *Joosia*, *Phitopis*, and *Stilpnophyllum*) sub endemic. Species endemism, on the other hand, is fairly high, *ca.* 59% for the tropical Andes as a whole. Similarities to non-Andean areas in species composition is slight, typically 10–20% to adjacent lowland areas and 1–5% to non adjacent areas. The low rate of generic endemism combined with a high rate of specific endemism and high species level distinctness may conceivably be interpreted as evidence of a strong adaptive radiation in geologically recent times. The Andean flora seems mainly to have arisen through adaptation of pre-existing lowland ancestors to montane conditions. This is probably the case with all members of the tribe Psychotrieae, which comprises *ca.* 43% of the total flora. Also the tribe Cinchoneae seem to be derived from a lowland tropical lineage. The Hedyotideae (12.5% of the flora) probably represent an ancient tropical montane element. Elements of tropical dry forest derivation (6.6%) and of Central American derivation (3.3%) have possibly arrived later. Elements, that appear to have invaded the Andes from temperate areas after the uplift, are few (4.8%), and may be restricted to the three members of the tribe Anthospermeae and, possibly, the genus *Galium*.

Lennart Andersson, Dept. of Systematic Botany, University of Göteborg, Carl Skottsbergs Gata 22, S-413 19 Göteborg, Sweden. Telephone 46-31-418700, Fax 46-31-823975.

LECTURE

Phytogeographical Relationships of Cloud Forest and Páramo on the Montanas de Guaramacal, Edo. Trujillo, Venezuela

G. Aymard C., L. J. Dorr and L. Barnett.

An inventory of the gymnosperm and angiosperm species found in cloud forest and páramo on the Montanas de Guaramacal serves as the basis of a study of the phytogeographic relations of this area, which is now preserved as the Parque Nacional "Cruz Carrillo." The park, which is situated at approximately 9°10–16'N, 70°11–15'W and covers 2000–3200 m in altitude, is at the northern end of the Cordillera de los Andes in Edo. Trujillo, Venezuela. The study is based on extensive field collections and comparison with specimens in several herbaria (MER, MO, NY, PORT, US, and VEN). To date, 503 taxa have been identified as occurring in the Montanas de Guaramacal. The major families represented are: Compositae (67 species), Orchidaceae (63), Piperaceae (47), Rubiaceae (35), Ericaceae (28), Melastomataceae (26), Begoniaceae (10), Bromeliaceae (9), and Araceae (8). These families account for 58% of the total number of species inventoried. The distributions of species encountered in the present inventory were scored as occurring in one of the following phytogeographic groups: 1) widespread in the Andes of South America; 2) restricted to northern South America (Ecuador, Colombia, and Venezuela); 3) restricted to Venezuela; or 4) restricted to the study area. Comparisons were also made to other floristic inventories in similar regions and we conclude that the cloud forests of Guaramacal have a different species composition than other Andean cloud forests, and, at the same time, show a strong affinity with the high-Andean flora of northern South America (Ecuador, Colombia, and Venezuela).

Gerardo Aymard C., UNELLEZ-Guanare, Programa de Recursos Naturales Renovables, Herbario Universitario (PORT), Mesa de Cavacas, Edo. Portuguesa 3323, Venezuela. Telephone and Fax 58 57519-249. Laurence J. Dorr, Department of Botany, NHB-166, Smithsonian Institution, Washington, DC 20560, USA. Telephone 202-633-9106, Fax 202-786-2563. Email: mnhbo059@sivm.si.edu. Lisa Barnett, Smithsonian Tropical Research Institute, MRC438, Smithsonian Institution, Washington, DC 20560, USA. Telephone 202-786-2817.

POSTER

Chimborazoa lachnocarpa, an Endangered (?) Species Endemic to Ecuador.

H. T. Beck

The monotypic genus Chimborazoa was recently described by Beck (1992). Chimborazoa lachnocarpa is an endemic species to mountainous regions (1200–2000 m) in the provinces Chimborazo and Cotopaxi, Ecuador. The species has only been collected four times in the last 133 years: Spruce, 1859; Hitchcock, 1923; Wiggins, 1944; and Cerón M., 1987. The areas where it has been collected (Huigra and Shasmay, Prov. Chimborazoa; Río Pilalo, Prov. Cotopaxi) are under intense human disturbance. Following publication of the new genus Chimborazoa, Prof. Patricio Mena and I visited the most recent collection site at the Río Pilalo, where fruiting material had been collected. We were unable to locate the plant, but we did confirm that the area is heavily disturbed by wood gathering, fire, and grazing. Only disturbed forest was observed in remnant patches along the river and small ravines. The other collection sites at and near Huigra are dry habitats mostly denuded of forest cover and heavily disturbed by goats (Luteyn, pers. comm.). Given the rarity of collections and the endemic distribution of C. lachnocarpa in apparently disturbed habitats, this species should be considered rare and endangered. Efforts by botanists to locate additional habitats and define the ecology, including dispersal, of this species are needed.

Hans T. Beck, Institute of Economic Botany, The New York Botanical Garden, Bronx, NY 10458-5126, USA. Telephone 718-817-8970, Fax 718-220-1029.

LECTURE

Diversification of Onagraceae in Neotropical Mountains

P. E. Berry

Of the 16 genera and 655 species of Onagraceae, eight genera and 160 species occur in South America. This paper analyzes how each group arrived and differentiated in the continent, emphasizing those that primarily radiated in tropical montane environments. Four genera (*Boisduvalia, Camissonia, Clarkia,* and *Gayophytum*) each have just a single endemic species in temperate South America, all of them recently arrived disjuncts from the main center of their groups in western North America. *Ludwigia* has its greatest sectional and species diversity in South America, with 49 of the 82 species in the genus, but it is largely restricted to swampy lowland areas, and only a few weedy species enter the higher Andes. *Epilobium* has 12 of its 162 species in South America. Only one group of four species occurs in the tropical Andes and is recently derived from North American ancestors, while the remaining species were recently derived from Australasian taxa. In South America, *Oenothera* consists of 52 species belonging to three different groups, which are the result of at least three separate introductions from North America. One group that consists of six species is restricted to the northern Andes, and a second has two species in the southern temperate zone. The third group has 42 species that have radiated extensively in the Central and Southern Andes, mainly through habitat specialization and hybridization in conjunction with the formation of complex structural heterozygotes. The genus that has radiated most extensively in neotropical mountains is *Fuchsia*. Unlike the other genera, it probably originated in the austral temperate zone. One section of nine species shows a classical disjunction between the temperate Andes (one species) and the mountains of south-eastern Brazil, where there are five distinct and very local endemics (mostly on mountain peaks), one wide-ranging species with three subspecies, and two closely related but locally differentiated species. The tropical Andes are inhabited by three sections: one monotypic and restricted to marginal forest habitats of the Pacific slopes of Peru; a second section with 16 mostly local cloud forest endemics from Venezuela to Bolivia; and the main section *Fuchsia* with over 60 species in cloud forests from Venezuela to Argentina. Further details of the speciation and distribution patterns of the Andean fuchsias will be presented.

Paul E. Berry, Missouri Botanical Garden, P.O. Box 299, St. Louis, MO 63166-0299, USA. Telephone 314-577-5186, Fax 314-577-9596, Email: berry@mobot.org.

POSTER

The Araliaceae in Ecuador

F. Borchsenius

The Araliaceae is a poorly known family in the Ecuadorean flora. It is estimated that about 65 species occur in Ecuador, belonging to three genera, *Dendropanax, Schefflera* (incl. *Didymopanax*), and *Oreopanax*. Of these *Dendropanax* is the least diversified with an estimated four species. These are trees and shrubs occurring in lowland forest; two species are so far identified. *Schefflera* is estimated to include about 25 species, most abundant in premontane and lower montane forest in the Andes; only two species have so far been identified with certainty. *Oreopanax* includes an estimated 35 species, mainly distributed in montane forest; seven species have so far been identified. The genus is taxonomically complicated due to serial variation in leaf morphology, and the number of diocious or polygamodioecious species in the genus. The present study was initiated in august 1992 with the aim of treating the Araliaceae for the Flora of Ecuador.

Finn Borchsenius, Herbario QCA, Departamento de Biologia Pontificia Universidad Católica del Ecuador Apartado 17-01-2184, Quito, Ecuador. Telephone 593-2-529-250 ext. 279, Fax 593-2-657117 or 593-2-466 919.

LECTURE

Native Plants, Human Communities and Conservation in Montane Forests in Peru

A. Brack-Egg

From Peru 25,000 species of plants are known; of these no less than 3140 species are being used for 33 different purposes. About 1005 species are cultivated and of these 128 may be considered domesticated. Of the native plants 682 are used for food (226 of them cultivated); 1044 species are medicinal; 55 species are used as fertilizer (52 cult.); 60 species are used for oil and fat (25 cult.); 292 are used in agroforestry; 64 species as antidotes (19 cult.); 46 species for perfumes and aroma (18 cult.); 25 species as spice (18 cult.); 444 species for wood and construction (75 cult.); 75 species as cosmetics (24 cult.); 68 species as stimulants, narcotics, psychotropics and hallucinogens (17 cult.); 47 species for veterinary purposes (10 cult.); 99 species for fibre (27 cult); 86 species for fodder (50 cult.); 34 species for magic purposes and shamanism (9 cult.); 553 species for ornamentals; 77 species for controlling human reproduction (32 cult.); 128 species for dye (36 cult.); 207 toxic species (35 cult.). Peru is maybe the country with most of its native species being used, cultivated, and domesticated. This represents a great richness of genetic resources which is only little studied and conserved; it is a great cultural richness based on thousands of years of tradition concerning uses and knowledge; it is an economic richness of which it is difficult to estimate the true value. A very important part of this richness may be found in the Andean montane forests, including wild relatives of domesticated species. At present there is a serious genetic erosion and, because of the process of aculturalisation, the original cultures are abandoning traditional use and production systems. The conservation and saving of the knowledge of native plant species is a most urgent undertaking. For the conservation of these resources the following activities should receive high priority: 1) conserve production systems and traditional uses of the aboriginal cultures; 2) increase ethnobotanical research and gathering of information about uses; 3) improve the condition of protected areas and increase their area; and 4) development projects with national and international funding should put priority on the use of native species and avoid the introduction of foreign species.

Antonio Brack-Egg, Programa de Biodiversidad de UNDP-GEF en el Tratado de Cooperación Amazónica, Avda. 10 de Agosto 3560 y M. de Jesús, Quito, Ecuador. Fax 593-2-434388.

LECTURE

Conservation of the Montane Subtropical Forest in North-western Argentina

A. D. Brown

In Argentina montane subtropical forests are found in the Atlantic forest in the province of Misiones (Northeast) and the Tucuman-Bolivian forest in the Yungas (Northwest). Yungas, located 300–2500 m above sea level, covers about 2.5 million hectares and has orographic rains ranging from 1000–3000 mm per year of which about 80% fall between November and April. In the Yungas, the vegetation of premontane areas are currently undergoing rapid and complete transformation whereas the areas at intermediate elevations still have an acceptable protection status. A conservation strategy for these forests should include the process of transformation of natural landscapes into productive landscapes. The strategy should take into account that the forest is important for regulation of hydric conditions and for the subsistence of human activities along the entire altitudinal gradient. The human activities include forest exploitation, shifting agriculture, and cattle management. An important aspect of the conservation strategy should be the maintenance of corridors between existing reserves and new ones to be created in key areas. To achieve this goal both biological diversity and stability should be considered. Using these criteria it appears that high priority should be given to the conservation of the area between Baritú and Calilegua National Parks, because of its great size (*ca.* 500,000 ha), high diversity (over 250 tree species alone), its importance for the hydric regulation of most of the high basin of the Bermejo river, and the persistence of the native populations which use the land intensively but in a way which is ecologically compatible with the conservation of the area. These populations have a profound knowledge of the plant resources of the area which have not been studied yet. This part of the montane subtropical forest continues into a section of the Bolivian territory. Both countries should therefore be involved in this conservation effort.

Alejandro Brown, Lab. Invest. Ecologicas de las Yungas, Univ. Tucuman, C.C 34 (4107) Yerba Buena, Tucuman, Argentina. Telephone 081-252806, Fax 081-252806.

LECTURE

Montane Species-distributions in Costa Rica; the Role of Altitudinal Differentiation in Generating Species Richness

W. Burger

The native species of Euphorbiaceae, Gramineae, Lauraceae, and Rubiaceae are used to examine patterns of elevational distribution in Costa Rica. While these families' distributions differ in interesting ways, they also show concordant patterns. The upper altitudinal limits of these *ca.* 900 species show no clear decline until the 2100–2400 m range where colder temperatures probably become limiting. There is no evidence for abrupt physical discontinuities or soil changes in the 200–2000 m range. Nevertheless, many of Costa Rica's species, both endemic and wide-ranging, have an upper or lower altitudinal limit in the benign 1000–1600 m range on the evergreen Caribbean slope. In addition, some parapatric closecongeners are sharply separated by their non-overlapping altitudinal ranges within this range. This raises the question: what actually determines elevational limits on this moist slope where temperatures are never very hot or very cold? It is hypothesized that phytopathogenic interactions may be an important determinant of sharp altitudinally defined species boundaries. While an uncommon occurrence, closely parapatric sisterspecies separated by their non-overlapping altitudinal range are found in many different families. These recently diverged species-pairs may be evidence for a process of pathogen-mediated montane speciation.

William C. Burger, Department of Botany, Field Museum of Natural History, Chicago, IL 60605-2406, USA. Telephone 312-922-9410, ext. 318.

LECTURE

A Biogeographical Approach to the Vascular Plant Flora from the High Mountain of the Farallones-de-Cali (W. Cordillera, Colombia)

E. Calderón-Sáenz

An inventory of the vascular plants of the Farallones-de-Cali national park of 160,000 ha was made to determine level of endemism and biogeographical affinity of this isolated mountain. Herbarium material from VALLE, COL, and CUVC and the literature was critically examined. The Farallones-de-Cali includes a steep and high segment of the Colombian Western Cordillera. Upper vegetation belts (2500–4200 m) are isolated from the rest of the Western Cordillera, which is considerably lower south and north of the Farallones. Deeply faulted basalts of Cretaceous age, with intercalations of marine sediments, characterize the upper parts of the Farallones; a steep and rocky páramo, 10 km long, with mostly primary soils, sometimes grade into peat bogs. Complex, stratified and well drained soils are lacking. Life forms include cushion plants, bunch grasses, dwarf shrubs, caulescent and acaulescent rosettes. Some typical floristic páramo elements of the Central and Western Cordillera are lacking, such as Espeletiinae, *Calceolaria, Aragoa, Polylepis,* and *Lupinus.* Endemics (specific and infraspecific) make up *ca.* 17% of all taxa so far listed. Compositae have most endemic taxa (10), although mostly infraspecific. Ericaceae have the largest number of endemic species (6 incl. 3 *Themistoclesia).* Other families rich in endemics are Aquifoliaceae (3 *Ilex),* Araliaceae (2 *Schefflera* and 1 *Oreopanax),* Piperaceae (3 *Peperomia),* and Melastomataceae (2 *Miconia,* 1 *Axinaea).* Brunneliaceae and Rubiaceae have two endemics each. Endemic species commonly found in this peculiar páramo include: *Ilex suprema, Puya occidentalis, Gaultheria oreogena, Themistoclesia compta, Macrocarpea duquei,* and *Diplostephium farallonense.* An interesting, disjunct population of *Castratella piloselloides* (Melastomataceae) occurs in the páramo of Farallones-de-Cali (the nearest populations of this species occur in the Eastern Cordillera of Colombia and in the Andes of Venezuela; it is absent from the Central Cordillera and from the rest of the Western Cordillera of Colombia!). Certain páramo elements of eastern-central Andean distribution that have already colonised the northern segment of the Western Cordillera (like *Aragoa* and *Espeletia),* have not reached the Farallones-de-Cali, yet. Close floristic affinities with southern Colombian páramos (Macizo Colombiano, Nudo de Los Pastos) are demonstrated by presence of *Odontoglossum compactum, Ilex pernervata,* and *Hedyosmum cumbalense,* reflecting active floristic exchange in the past.

Eduardo Calderón-Sáenz, FUNDACION FARALLONES: Cra. 24B No. 2A-99, Miraflores, Ap. Aéreo 20803, Cali, Colombia. Telephone 568335.

LECTURE

Contributions to the Phanerogam Flora of Manu National Park, Cusco, Peru

A. Cano, B. León, K. R. Young and R. B. Foster

The montane and high-Andean flora of Manu National Park in Peru is still little known, despite the great scientific interest for the park. For this reason we have carried out floristic work in these regions, and until now we have made over 4000 collections. This presentation concentrates on the areas of Acjanaco, Tres Cruces, and Pillahuata, lying at elevations of 2400–3600 m. Based on our collections, and others from various herbaria, we have found over 100 families of phanerogams, 240 genera and about 500 species. The most species rich families are Asteraceae, Ericaceae, Poaceae, Orchidaceae, Rosaceae, Melastomataceae, and Solanaceae.

Asunción Cano and Blanca León, Museo de Historia Natural, Av. Arenales 1256, Apartado 14-0434, Lima 14, Peru. Kenneth R. Young, Department of Geography, University of Maryland, Baltimore County Campus, Baltimore, MD 21228, USA. Robin B. Foster, Smithsonian Tropical Research Institute, Aptdo. 2072, Balboa, Republic of Panama; Department of Botany, Field Museum of Natural History, Chicago, IL 60605, USA.

LECTURE

The Orchid Flora of the Table Mountains (Tepuis) of the Guayana Highlands (Pantepui): Diversity and Phytogeography

G. Carnevali

The Guayana Highlands are defined in this paper as all areas in the combined Venezuelan Guayana/Guianas/N. Brazil region at elevations above 1000 m, and conforming the floristic province of Pantepui. Of the ca. 1000 species of orchids in the Guianas area, 349 occur in Pantepui, ca. 230 reach altitudes of 1500 m while ca. 130 occur above 2000 m. Endemism levels are 32.9%, 37.5%, and 53.0% for taxa occurring between 1000–1500 m, 1500–2000 m, and above 2000 m, respectively. For species only occurring above 1500 m and 2000 m, the endemism levels are higher (58.0% and 69%, respectively). The largest orchid genera in Pantepui are *Epidendrum*, *Plaurothallis*, *Mazillaria*, and *Octomeria*. Phytogeographical correlates of orchid diversity are discussed for factors such as altitude, precipitation, tepui area, tepui isolation, and habitat. Most of the endemisms among the Pantepui orchids are associated with the unique habitats of the tepuis. Two major phytogeographical areas and possible centers of diversification can be recognized for orchids in Pantepui, the North-eastern Tepuis and the South-western Tepuis, the first area being more diverse and rich in endemisms. Autochtonous, Andean, West Indian-Central America, South Brazilian, and Guianan-Amazonian floristic elements are recognized among the orchidaceae of Pantepui. It is postulated that the climatic changes of the Pleistocene that seem to have affected the tepui summits strongly, contributed to much extinction on the summits of the tepuis and to the evolution of new taxa allopatrically and by speciation, probably via genetic transilience associated with recolonisation and habitat specialization.

Germán Carnevali, Fundación Instituto Botánico de Venezuela, Herbario Nacional de Venezuela (VEN), INPARQUES, Aptdo. 2156, Caracas 1010-A, Venezuela; Current Address: Missouri Botanical Garden, P.O. Box 299. St. Louis, Missouri 63166-0299, USA.

LECTURE

Deforestation of Montane Rain Forests in Colombia as a Result of Illegal Plantations of Opium (Papaver somniferum)

J. Cavelier

Opium was introduced in Colombia in the 1980's. Since then, there has been a steady increase in the areas converted from primary montane rain forests to rudimentary agricultural fields for the cultivation of opium and the later production of morphine and heroine. The forests are cleared using traditional slash-and-burn agriculture techniques. Although these activities are all illegal, poppy fields are now widely spread in remote mountainous areas along the Central and Eastern Cordilleras of the Colombian Andes. Plantations have been located in more than 100 counties in 13 of the 16 Departments (provinces) that have territories on the Cordillera of the Andes. During 1992 alone, a total of 11,000 ha (11 km^2) of primary montane rain forests were cleared to open fields for the cultivation of opium. Ninety percent (90%) of the deforestation took place in lower montane rain forests (1000–2500 m) with the remaining ten percent (10%) in upper montane rain forests (2500–3500 m). Although deforestation of montane forests is limited when compared to deforestation of lowland moist forests (6500 km^2/year, Myers, 1989, Deforestation Rates in Tropical Forests and their Climatic Implications, Fiends of the Earth, London), the present extent of primary montane rain forests is also much smaller. Thus, these ecosystems are much more likely to be destroyed and disappear in the near future. A map of the original and present areas covered by montane rain forests, and the location of active deforestation, is being prepared with the aid of air photography, satellite images, and information provided by the antinarcotics police in Colombia. This institution is fighting the destruction of these unique ecosystems by fumigating poppy fields with "Roundup". During 1992, a total of 9000 ha of opium were destroyed. No effects of the fumigation on the forest-edges have been reported. The conservation of the flora and fauna of the remaining montane forests depends largely on the success of the government to control the cultivation of this plant. Montane forests in national parks are not threatened for the moment and seem to be the only areas that will be saved in the future. Colombia is in great need of maintaining and creating new national parks along the Andes to ensure the conservation of a unique flora and fauna that is still largely unknown.

Jaime Cavelier, Departamento de Ciencias Biologicas, Universidad de los Andes, AA 4976, Santafe de Bogota, Colombia. Telephone 57-1-2849911, Fax 2841890, Email: jcavelie@andescol.

LECTURE

The Puna and the Development of Central Peru

E. Cerrate

One of the most promising zones for central Peru is the Puna, located at elevations of 3800–5000 m, with soils on slightly sloping terrain, a cold and dry climate and abundant precipitation. Because of its geomorphology this area was the cradle of the primitive agrarian societies in Peru. Because of its exceptional flora and fauna, its extraordinary and beautiful scenery, abundance of liquid as well as frozen water, and few diseases, the most difficult obstacle to human life is the scarcity of oxygen. This problem is overcome through the use of naturally dried leaves of the "coca" plant (*Erythroxylon coca*), which grows in the forests of the area. The Puna is destined to be the future shelter for healthy Peruvians, physically and mentally strong, and capable of carrying out the progress and development of Peru.

Emma Cerrate, Universidad Nacional Mayor de San Marcos, Museo Historia Natural, Av. Arenales 1256, Apartado 14-0434, Lima, Peru. Telephone 710117.

LECTURE

Ecology and Management of Vegetation in the Montane Forest Belt in Costa Rica

A. Chaverri

In spite of its small size (ca. 51,000 km^2) Costa Rica has a high biodiversity. Most natural ecosystems containing this biodiversity are represented in national parks or biological reserves, montane and lower montane forests being well represented. Increasing pressure on the montane forests outside protected areas makes it is necessary to start a sustainable management, in order to conserve them. This paper is especially concerned with the oak forests in Costa Rica at elevations of 2300–2900 m, on the Pacific slope of the Cordillera de Talamanca, NW of Cerro de la Muerte. A detailed study of land capacity outside protected areas, separating forest protection from forest production land, must be made. Within forest production areas, ecological and silvicultural research, which will give important information for the management plan, is necessary. Detailed studies on vegetation structure and floristic composition, and a forest inventory, will provide information (frequency, density, basal area, diametric distribution, and natural regeneration) about the forest. Studies of phenology, growth, demography and dynamics provide information needed for managing the forests. Four main undisturbed forest communities were located in the montane belt from 2200–3100 m: *Myrsine pittieri-Quercus costaricensis, Q. costaricensis-Q. copeyensis, Geonoma hoffmanniana-Q. copeyensis, and Q. seemannii-Q. copeyensis.* In the montane forests, oaks represent 35–79.7% of all trees and their basal area varies from 28–70 m^2/ha. The most abundant and frequent species in the montane forest belt, *Q. copeyensis,* has the widest J-shaped diametrical distribution, reaching 140–149 cm dbh. Natural regeneration is high, often above 5000 saplings per hectare. The intrinsic characteristics of the forest: low woody plant diversity, higher commercial basal area than in lowland forests, high natural regeneration, plus the fact that oaks grow well in small and middle-sized gaps, make sustainable management of these forests feasible. The slow growth of most montane forest trees would implicate silvicultural treatments farther spaced in time from each other. Management plans should include seed trees and "ecological" trees in the forest, and ecologically sound harvest by skilled operators. A selective system which maintain the natural age structure (excl. over-sized trees) may be implemented. Sustained management will conserve forests outside protected areas, in the montane forest belt, and elsewhere in the tropics. Forest management should be accompanied by reforestation with native species, natural revegetation, and the promotion of agroforestry systems.

Adelaida Chaverri-Polini, Escuela Ciencias Ambientales, Universidad Nacional, Heredia 3000, Costa Rica. Telephone 506-376363 ext. 2291 or 2293, Fax 506-377036 or 506-380086-2291 or 2293.

LECTURE

Diversity and Biogeography of the Neotropical Andean Mosses

S. P. Churchill, D. Griffin III and M. Lewis

The neotropical Andes, for its size, represents one of the highest diversity levels for mosses in the world. Based on a preliminary list for the neotropical Andean countries (Venezuela, Colombia, Ecuador, Peru, and Bolivia) there are approximately 2080 moss species distributed among 340 genera and 80 families. Taxonomic inflation, due to excessive naming by Europeans in the latter part of the 19th century and early part of this century, has contributed some 300–400 names to the figure given here that will likely prove to be synonyms. However, the estimates and percentages given here should balance out since this applies to the Neotropics in general. The five largest families, accounting for just over 50% of the total number of species, are the Pottiaceae, Bryaceae, Dicranaceae, Orthotrichaceae, and Callicostaceae. Speciose genera, with 30 or more species, include among others: *Breutelia, Bryum, Campylopus, Didymodon, Fissidens, Lepidopilum, Macromitrium, Schizymenium,* and *Sematophyllum.* Elevational data for the neotropical Andean countries suggests that approximately 16% of the species are known to occur at or below 1000 m, and of that only 7% are unique to the lowlands, the remaining 9% also occur in the highlands. Further-more, about half of the 7% are associated with the lowlands adjacent to the Andes and do not extend in any significant way into the Amazon basin. The Andean region thus contains 93% (*ca.* 1915 species) of the total moss diversity for the five Andean countries, and is estimated to be 7.5 times richer than the entire Amazon basin. Montane forest mosses contribute between 30–40% of the total species richness of the Andean region, the majority of these occur as epiphytes, only a relatively small number are found on soil, humus or logs. The remaining Andean species, about 60–70%, are found in open montane (natural forest margins and borders, clearings, landslides, and disturbances related to human occupation) and páramo/puna regions. Geographical patterns suggest that the wetter and more extensive montane forests and borders of the northern Andes may be richer than in the central Andes. Conversely, the drier central Andes are possibly more species rich in the open montane and puna regions, at least for certain families (*e.g.,* Grimmiaceae, Pottiaceae). Geographical affinities largely follow patterns previously suggested for Andean plants, a major neotropical element followed by Andean accounting for *ca.* 70–80% of the distributional ranges. Holarctic-North Temperate and Austral-Antarctic/South Temperate elements are nearly equal, *ca.* 2–4%, and occur primarily at high elevations. Taxonomic and floristic progress among the Andean countries has increased exponentially in the last 20 years. There are now a total of some 240 scientific papers treating the taxonomy of mosses at some level for the Andes. Among the five Andean countries, early exploration accounts with many new species described, recent floristic studies, and related papers on mosses, are estimated as follows: Colombia -60, Venezuela -45, Ecuador -35, Peru -35, and Bolivia -20. This also demonstrates a major dilemma for Latin Americans, few of the some 430 publications are available in these countries. Monographic studies are essential, however floristic treatments, *i.e.*, floras and florulas are equally important to promote a resolution of not only taxonomic problems, but to commence investigations of all aspects of comparative biology with regards to mosses.

Steven P. Churchill, Institute of Systematic Botany, New York Botanical Garden, Bronx, N.Y. 10458, USA. Telephone 718-817-8641, Fax 718-562-6780. Dana Griffin III, Florida State Museum, Museum Road, University of Flora, Gainesville, FL 32611, USA. Telephone 904-392-6577, Fax 904-392-8783. Marko Lewis, Herbario Nacional de Bolivia, Correo Central Cajon Postal 10077, La Paz, Bolivia. Telephone 591-2-792582, Fax 591-2-359-491.

LECTURE

Diversity and Distribution of the Andean Woody Bamboos (Poaceae: Bambuseae)

L. G. Clark

Woody bamboos are a common, sometimes dominant element in neotropical montane forests, particularly in secondary vegetation. Of the 22 genera and approximately 410 species of woody bamboos in the New World, seven genera and an estimated 130 species (32%) occur in the central and northern Andes. None of the genera is exclusive to the Andes, but approximately 90% of the species are endemic. The seven Andean genera are *Arthrostylidium, Aulonemia, Chusquea, Elytrostachys, Guadua, Neurolepis,* and *Rhipidocladum;* of these, *Aulonemia, Chusquea, Neurolepis,* and *Rhipidocladum* have the majority of their diversity in the Andes and are considered to be Andean-centered. The 10 Andean species of *Rhipidocladum* range from 450–2900 m in elevation and are associated exclusively with montane forests. The two species of *Rhipidocladum* sect. *Didymogonyx* can form extensive stands in upper montane forest. *Aulonemia* is characteristic of the upper montane forest/subpáramo/páramo zones; the 15 Andean species usually occur at elevations above 1800 m. Although a few species can form extensive stands, most species tend to occur as scattered, discrete clumps. *Neurolepis,* a poorly understood genus, encompasses an estimated 20 species with the diversity apparently centered in Ecuador. The majority of its species occur in the subpáramos and páramos of the Andes, but a few species, *e.g., N. pittieri,* are found in upper montane forests. *Chusquea,* with an estimated 180 species, is the most diverse woody bamboo genus in the world. Approximately 75 of its species (42%) occur in the central and northern Andes, and 71 of them (95%) are endemic. The seven currently recognized sections of *Chusquea* are all represented in the central and northern Andes; *Chusquea* sect. *Longiprophyllae* is restricted to this area. The shrubby species (*Chusquea* sect. *Swallenochloa)* are characteristic of the subpáramo/páramo zone, whereas the viny, scandent, or large and erect species of the other sections occur in the montane forests usually at elevations above 1000 m. A number of species are invasive and weedy, and their abundance and distribution have almost certainly been affected by both natural and man-made disturbances.

Lynn G. Clark, Department of Botany, Iowa State University, Ames, IA 50011-1020, USA. Telephone 515-294-8218, Fax 515-294-1337.

LECTURE

Species Diversity at Montane and Lowland Sites in Colombia and Ecuador

T. Croat

The purpose of the study was to compare the relative size of different genera in specific montane and lowland sites in western South America to determine how species diversity of the Araceae is related to life zone and elevation. The sites selected were those sufficiently well collected that most species occurring in the area were likely to be represented in the sample and large enough to represent most species occurring at a particular elevation and life zone. Specific montane sites selected were the Reserva Natural Río Guajalito and Reserva ENDESA in Ecuador (both in Pichincha Province) and Reserva Natural La Planada in Colombia (Nariño Province). Specific lowland sites with which they will be compared are Reserva Jatun Sacha in Napo Province of Ecuador and the Bajo Calima region of the lower Río Calima in Colombia. Thus study sites include lowland sites on both the Pacific slope of NW South America as well as in the Amazon basin. These studies have shown that species diversity is greatest in very wet lowland areas and that species diversity decreases with increases in elevation especially above about 1500 m. However, some genera such as *Anthurium* are shown to diminish as a percentage of the total species as one approaches the lowlands and others such as *Philodendron* diminishes as a percentage of the total as one approaches the montane habitats. Certain genera such as *Spathiphyllum* and *Dieffenbachia* may be altogether absent from the upland sites while others such as *Stenospermation* may be equally abundant in wet lowland sites and in montane sites. While Araceae appear to be the dominant epiphytic group in both the lowland and montane sites, montane forests have proportionately more individuals of particular species per unit area than is true at the lowland sites. Montane species are in general more poorly known taxonomically than lowland Amazonian sites, but are substantially more well known than lowland Pacific coastal sites.

T. Croat, Missouri Botanical Garden, St Louis, MO 63166-0299, USA.

LECTURE

Neotropical Montane Forests: A fragile Home of Genetic Resources of New World Crops

D. G. Debouck and D. Libreros Ferla

This paper briefly presents the different kinds of Neotropical montane forest and their location in Latin America, as home of genetic resources of New World crops (*i.e., Carica, Cyphomandra, Lupinus, Lycopersicon, Solanum, Pachyrhizus, Passiflora, Persea, Phaseolus, Polymnia, Prunus, Rubus*). Some biological attributes relevant to plant conservation are also presented. It reviews plant ancestry relationships for a certain number of these crops and presents evidence about the structure of their gene pools. The significance of these results for the integrated conservation using methodologies *ex situ* and *in situ* is discussed. After reviewing the protection status of these gene pools and their associated vegetation, it concludes to the possible actions that have to be launched now for their conservation.

Daniel G. Debouck, International Board for Plant Genetic Resources, Group for the Americas, c/o CIAT, Apartado Aéreo 6713, Cali, Colombia. Telephone 57-23-675050, Fax 57-23-647243. Dimary Libreros Ferla, International Board for Plant Genetic Resources, Group for the Americas, c/o CIAT, Apartado Aéreo 6713, Cali, Colombia. Telephone 57-23-675050, Fax 57-23-647243.

POSTER

Tree Vegetation Structure of two Montane Forest Sites in The Southern Venezuelan Guayana as Influenced by Soil Characteristics, Topography and Geomorphological Variation

L. Delgado

Vegetation studies in tropical forest systems have been underway in the Venezuelan Guayana since 1985 by the Inventory of Natural Resources Project of the CVG-TECMIN of Venezuela. Proceeding systematically using 1:250.000 radar and satellite images, representative study sites were pinpointed throughout the States of Delta Amacuro, Bolívar and Amazonas for vegetation structural analysis in relation to soil type and geomorphological composition. As part of these studies, two 0.5 ha vegetation sampling plots were established in a rolling plateau area between 1450 m and 1600 m in the Maigualida mountain range in the border area between Bolívar and Amazonas states of southern Venezuela. Floristic and forest structure studies were undertaken in the plots. In addition, such soil physical-chemical characteristics as texture, pH, Mo, CIC, *etc.*, as well as relief type, slope and geological aspects were determined. The forest tree species importance value index was calculated utilizing frequency, abundance and dominance data. Geological analysis of exposed rock showed a dominance of granite. In both study sites, the soil types were ultisols. However, in the lower site at 1450 m, the soil texture was determined as [franco] near the surface and clay at the greater depths sampled, with an overall pH range of 3.2—5.6. In the 1450 m site, the forest canopy height oscillated between 24—30 m, and the most important species were: *Pouteria* sp. (16%), *Licania* sp. (11%), and *Protium* sp: (11%). At the 1600 m study site, the soil texture was [arenoso francosa] near the surface and sandy clay at greater depths, with a pH range of 4.1—4.9. The forest canopy height was lower at a maximum of 22 m, and the most important species were: *Pouteria* sp. (33%), *Macrolobium* sp. (11%), and *Aidina* sp. (7%). Forest structure and floristic composition within the two study areas were most influenced by slope, soil substrate texture and geomorphological variation.

Luz Delgado, Venezuelan Guayana Natural Resources Project, CVG-TECMIN, Edif. CVG, Av. Germania, Piso 2, Ciudad Bolívar, Venezuela.

LECTURE

Forest Types of Cerro Duida, Venezuelan Guayana

N. Dezzeo

Cerro Duida, a huge mountain system of approx. 1000 km^2 and a maximum elevation of 2375 m, is located in the central Amazonas State of southern Venezuela and belongs to the biogeographical province of Pantepui. Together with Cerro Marahuaka (2800 m) and Cerro Huachamakari (1950 m), Cerro Duida forms part of the "Duida-Marahuaca" National Park; the present results were obtained during the elaboration of a detailed vegetation map (1:100,000) for the future management plan of this national park. The main body of Cerro Duida is made up of a large sandstone meseta inclined from south (2375 m) to north (*ca.* 1100 m), with an extensive river system draining the gradually ascending rocky plateaus. The climate is assumed to be almost non-seasonal, with high average annual rainfall (3000–4000 mm), and a variable temperature regime ranging from hot, macrothermic at the base (150 m) to mesothermic (12–15°C average annual air temperature) in the summit region. Soils vary from deep mineral soils in the northern section to extensive organic soils (peat) in the higher southern section of the mountain. By far the largest part of Cerro Duida is covered by an intricate mosaic of forest types, of which the following are the most important ones: 1. Macrothermic basimontane forests, extending along the lower slopes of the massif up to *ca.* 500 m; evergreen, medium to tall, dense forests with very high species richness. 2. Submesothermic lower montane forests, extending widely between 500 and 1400 m, both on the upper slopes, as well as in the interior plateaus of the massif. Two sub-types are readily distinguished: (a) evergreen, medium to tall, dense forests growing on sandstone and igneous or meta-sedimentary substrate; the dominant trees in these species-rich forests are *Dimorphandra* sp. (Caesalpiniaceae) and *Perissocarpa umbellifera* (Ochnaceae). (b) evergreen, sclerophyllous, medium to low and relatively open forests growing exclusively on sandstone; the dominant species are *Neotatea longifolia* (Theaceae) and *Tyleria grandiflora* (Ochnaceae). 3. Mesothermic montane forests. These low, open, sclerophyllous forests cover part of the higher southern plateaus between 1400 and 1800 m and are growing on peat. Their tree flora is relatively poor, consisting mainly of *Bonnetia crassa* and *Neotatea longifolia* (Theaceae), *Tyleria* sp., and *Gleasonia duidana* (Rubiaceae).

Nelda Dezzeo, CVG-EDELCA, Dept. Estudios Básicos, Torre Maxy's Edif. C., Piso 8. Puerto Ordaz, Edo. Bolívar, Venezuela. Telephone 086-603784, 603721, Fax 58-86-603-554.

LECTURE

Floristic Inventory and Biogeographic Analysis of Montane Forests in Northern Peru

M. O. Dillon, A. Sagástegui A., I. Sánchez V. and S. Llatas Q.

While attention has been focused on the plight of tropical rain forests of the Amazonian Basin, the biologically rich montane forests associated with the Andean Cordillera continue to be destroyed at a staggering rate. The destruction of forests during the last 50 years in north-western Peru has not been documented. In an attempt to better understand the floristic diversity of northern Peru's remaining montane forests, the Departments of Cajamarca and Piura have been extensively sampled during the last five years. The floras of these areas represent the largest remaining tracts of relatively undisturbed montane forests east of the Río Marañón (78°W) and south of the Huancabamba deflection in Piura (4°30'S). The project has established a relational, specimen-oriented database consisting over 3000 records with locality and habitat data from project collections and herbarium surveys. Floristic data from this inventory were examined for biogeographic patterns and used to test hypotheses of fragmentation or dispersal origins. Northern Peru possesses a mosaic of environments containing high levels of endemism and diversity, including many species new to science. It appears likely that these remnant forests represent the last of a once more continuous expanse of forest extending throughout the northern and central Andean Cordillera. In response to glacial cycles and world-wide sea-level lowering, the climate of northern Peru was wetter and cooler as recently as 18,000 years ago. With the retreat of the glaciers and the re-establishment of the cool Humboldt Current, a warmer, drier coastal climate developed from *ca.* 12°S latitude (near Lima) and north to southern Ecuador. These climatic changes and continued uplift of the Andean Cordillera have resulted in a decrease in precipitation and therefore suitable habitat. The remaining forests can be viewed as islands of vegetation which now have their nearest source populations north of the Huancabamba deflection and show strongest connections with moist forests of central and southern Ecuador. They also share taxa with forests east of the Río Marañón and the "ceja de la montaña" of the eastern versant. Several species have their southernmost distributional terminus in the forests of north-western Peru. It is hopeful that baseline data derived from these inventories will allow for recognition of regions in need of conservation attention and help to measure the extent of future changes associated with human encroachment.

Michael O. Dillon, Department of Botany, Roosevelt Road at Lake Shore Drive, Chicago, Illinois 60605-2496, USA. Telephone 312-922-9410. Abundio Sagástegui A., Universidad Antenor Orrego, Trujillo, Peru. Isidoro Sánchez V., Universidad Nacional de Cajamarca, Cajamarca, Peru. Santos Llatas Q., Universidad Nacional de Lambayeque, Lambayeque, Peru.

LECTURE

Monitoring Degradation of Montane Forests in the Andes: Two Case Studies

F. R. Echavarria

This paper describes two case studies in which land cover changes are quantified utilizing remote sensing techniques. The first case study was located in Peru's Huallaga Valley, where pre-montane forests (*ca.* 900 m elevation) have been cleared for coca cultivation. The study was limited to an area of approximately 1000 km^2 between the towns of Tingo Maria and Tocache Nuevo. The methodology used was both simple and appropriate for under-funded government agencies in the developing world on whose shoulders rests the heavy responsibility of management and conservation of these vanishing resources. This method utilized 1963 aerial photographs at a scale of 1:60,000 and 1976 satellite imagery-based, planimetric maps at a scale of 1:250,000. The second case study took place in southern Ecuador inside the provinces of Loja and Zamora/Chinchipe on the eastern side of Podocarpus National Park. In this study, more sophisticated techniques are used to monitor the clearing of montane forests for pasture, subsistence crops, roads, and mining operations. This study combines both remotely sensed data and *in situ* data. Satellite information from the Landsat MSS sensor was used to estimate deforestation rates at a regional scale between 1980 and 1986, while higher spatial resolution data from the Landsat TM sensor was used to model the topographic distortions of satellite data in montane topography. This modelling was done in a smaller area limited to the Bombuscaro watershed covered by montane forests ranging in elevations of 900–1400 m. The *in situ* data was mostly detailed information collected by botanists and biogeographers during extensive field work and reconnaissance in the region. These case studies, one in Peru and the other in southern Ecuador, provide examples of remote sensing techniques which can be used to more effectively monitor the degradation of montane forest in the Andes. The techniques used here provide valuable, large scale, synoptic information not available from other sources. This information can be extremely useful to botanists, biologists and ecologists attempting to identify forests that are under severe threat and forests worthy of immediate preservation due to their unique biological value.

Fernando R. Echavarria, Geography Department, University of South Carolina, Columbia, SC 29208, USA. Telephone 803-777-2942, Fax 803-777-4972, Email: echavar@kirk.geog.scarolina.edu.

LECTURE

Neotropical Montane Passifloraceae

L. K. Escobar

Passiflora, the largest genus in the Passifloraceae family, with approximately 500 species, is almost exclusively New World and Neotropical in distribution. The northernmost occurrence of the genus is the United States with 16 species, the southernmost is Argentina with 15. The greatest species diversity is found in Colombia (135 species) followed by Brazil with 114 species. But the greatest density of species, occurs in Costa Rica (47 spp.), Guatemala (54 spp), and Panama (34 spp), followed by the Andean nations of Ecuador (78 species) and Colombia. In the species dense countries of Central America, half of the species are found in habitats above 1000 m altitude and in South America, more than half are Andean, even though these habitats cover only a small fraction of the land area of the countries/ continent. Within the Andes above 1000 m altitude, species are concentrated in wet and moist areas. These are the western slopes of the Western Cordillera of Colombia and Ecuador and the eastern slopes of the Peruvian and Bolivian Andes. Only a few, wide-spread species are found in the dry, inter-Andean valleys. The division of the Colombian Andes into three Cordilleras at the Nudo de Pasto, provides many isolated habitats for genetic drift and divergence of genomes. The Sierra Nevada de Santa Marta, which is not part of the Andean chain, rises to more than 5000 m and has also provided habitats for the speciation of passion flowers. An analysis of endemic taxa shows that there are again, more subgenera and more endemic species and subgenera in Brazil and Colombia than in other geographic areas, but a greater density of endemic species and subgenera in Colombia. All of the subgenera endemic to Colombia are characterized by hummingbird pollination. Thus, speciation has proceeded by the spacial separation of taxa provided by the uplifting of the Andes, and ecological separation of taxa by specialized pollination.

Linda K. Escobar, Departamento de Biología, Universidad de Puerto Rico, Río Piedras, Puerto Rico.

POSTER

A Comparison of Moist and Wet Montane Forests in the Department of Antioquia, Colombia

L. K. Escobar, A. Uribe de Camargo and J. Vallejo Cossio

Two forests, located in the Department of Antioquia in northern Colombia, are compared structurally and floristically. One, classified as a pluvial forest in the Holdridge system, is located in the Municipio of Guatapé at 1850 m altitude on the eastern slope of the Central Cordillera. This forest receives more than 6000 mm of rain annually, a sum which increased by 1000 mm a year after the construction of the El Peñol reservoir in 1970. Peaks of rainfall occur in the months of April to May and August to October. Located on steeply sloping terrain, the soils of the forest are very acid (pH 3.7-4.2) and frequently waterlogged. The forest is dominated by palms, particularly *Dictyocaryum platysepalum,* a tree which reaches heights of more than 20 meters, and *Wettinia fascicularis,* a smaller understory species. This forest is floristically more similar to forests found at lower elevations, and is characterized by the many species of epiphytes, particularly of Araceae and Cyclanthaceae. Although frequently disturbed by falling *Dictyocaryum* leaves and tree trunks, several species of flowering plants new to science were discovered during the course of the study and few "weedy" plants were found there. The second forest, found in the interior of the Central Cordillera near Medellín, and also located on steep slopes at 2420–2610 m, receives 2000–3000 mm of rainfall annually, and is classified as very humid lower montane forest in the Holdridge system. Located along a road which passes by a few abandoned homes and a *Cupressus* plantation, this forest has suffered more disturbance than that of Guatapé, but still is characterized by several new, very endemic or uncommon species characteristic of the "Oriente Antioqueño", such as *Meliosma lindae, Meriania antioquiensis* and *Asplundia sarmentosa.* There are many herbaceous and woody climbers and epiphytes, particularly of orchids and aroids. The most common and diverse families of flowering plants are Rubiaceae, Solanaceae, Melastomataceae, Ericaceae, and Asteraceae. Both forests are privately owned and are under pressure by the residents of the area who cut trees for fire wood and tree ferns, small palms, orchids and mosses to sell.

Linda K. Escobar, Departamento de Biología, Universidad de Puerto Rico, Río Piedras, Puerto Rico. Alicia Uribe de Camargo and Javier Vallejo Cossio, Departamento de Biología, Universidad de Antioquia, Medellín, Colombia.

LECTURE

Floristic Evaluation of the Upper Marañón Valley

R. Ferreyra H.

The plant community and the most conspicuous indicator species of the vegetation of the upper Marañon valley are described. Among the floristically interesting new data are two new endemic genera: *Arnaldoa* and *Ferreyranthus*. Some undescribed species from the area will be treated.

Ramon Ferreyra H., Universidad Nacional Mayor de San Marcos, Museo de Historia Natural, Av. Arenales 1256, Apartado 14-0434, Lima, Peru. Telephone 710117.

LECTURE

Geographical Patterns of Neoendemic and Relict Species of Andean Forests Birds

J. Fjeldså

Patterns of diversification of montane forest birds in Ecuador, Peru, and Bolivia were analyzed by overlaying distributions of restricted range species which are part of young (Pleistocene) vicariance patterns and those of species showing relict patterns of various evolutionary age. While the total species richness is fairly uniformly distributed along the eastern Andean slopes, endemic and disjunct species show distinctive patterns. Neo-endemic density peaks which also coincide with peak numbers of old relict species are found along the continental divide in southern Ecuador, in Cordillera Colán, across the gap in the montane forest formed by the upper Huallaga River, in Cuzco, and in the Yungas of Cochabamba; there are additional peaks in peripheral areas with exclusive and precisely defined habitat in Cordillera Blanca, in Apurímac, and in the Boliviana/Tucumano forest. For 28 well studied areas along the eastern Andean slope, the number of neoendemic species is correlated with the number of relict species, but neither of these numbers are correlated with total species richness. The pattern suggests that vicariance events were less associated with physical barriers than with isolation in areas which were ecologically stable while the ecological regime of the region fluctuated. Ecological stability appears to be characterized by persistent rainfall, cloud-cover or mist caused by rather permanent atmospheric inversions. Patterns of endemism in montane areas may signal ecosystem functions of far reaching importance.

Jon Fjelså, Zoologisk Museum, Københavns Universitet, Universitetsparken 15, DK-2100 København K., Denmark. Telephone 45-35-321-000, Fax 45-35-3210-010.

LECTURE

The Genus Ribes (Saxifragaceae) in Ecuador

A. Freire-Fierro

Ribes is a large genus belonging to Ribesioideae of the Saxifragaceae sensu Engler. After the treatment by Janczewski (1905, 1906 and 1907), it has not been studied again. The genus is now being studied for "Flora of Ecuador." The taxonomic work is based on field work in Ecuador and literature. All Ecuadorean species of *Ribes* are functionally dioecius and belong to the subgenus *Parrilla*, which has rudimentary ovary in the staminate flowers and rudimentary stamens in the pistillate flowers. The genus has approximately eight species in Ecuador, and until now, the following species have been identified: *Ribes andicola* Jancz., *R. cuneifolium* Ruiz & Pavón, *R. ecuadorense* Jancz., *R. hirtum* H. & B., and *R. Lehmannii* Jancz. Among the characters used for the distinction of species are: the growth habit, size, shape and surface of the leaves, size and disposition of the inflorescence and the flower shape. The species with the greatest range in altitude and area is *Ribes hirtum*, which occurs at 2750–4200 m, from the province of Carchi to Loja, and in Sucumbíos, Napo and Morona Santiago; *Ribes cuneifolium* has the narrowest range, occurring at 3550–4150 m, and was collected only in Tungurahua, Azuay, and Napo.

Alina Freire-Fierro, Herbario QCA Departamento de Ciencias Biológicas, Pontificia Universidad Católica del Ecuador, Apartado 17-01-2184, Quito, Ecuador. Telephone 593 2-529270 ext. 279, Fax 593 2-567117.

Neotropical Montane Araliaceae: An Armchair Overview

D. B. Frodin

Araliaceae may not be as speciose as some families in tropicmontane forests, but here and there are conspicuous in form and/or mass; examples include *Harmsiopanax* in Malesia, *Oreopanax* in the Americas, and *Schefflera* over much of the tropics. Some are important colonisers of gaps and edges and so attract attention, such as *Oreopanax oroyanus* in Peru and *Harmsiopanax harmsii* in New Guinea. A few are even dominant or co-dominant and so enter ecological literature: such is *Schefflera mannii* in the islands of the Gulf of Guinea. Widely distributed species are, however, numerically a minority in the montane and upper montane forests of the neotropic, and no species occurs throughout. Evidence from *Schefflera* (now inclusive *Didymopanax*) and limited study of *Oreopanax*, *Dendropanax* and *Aralia* suggests that several centers of development can be recognized; these, and the patterns of diversity exhibited, will be discussed. The strongest discontinuity is between the Guayana and the Andean cordilleras; in *Schefflera* there is almost no overlap, even at strongly supraspecific levels, *Dendropanax* has only 1–2 endemic species (both below 1000 m) and *Oreopanax* in a meaningful way is absent. Other centers outside the main cordillera system include south-eastern Brazil, the Planalto (with in *Schefflera* a slight link with Guayana), Mexico (effectively without *Schefflera*) and a 'ring' centered on the Carribean Sea. In the Andes are a number of secondary centeres, some exhibiting recent diversification comparable to that in central New Guinea, where moreover in *Schefflera* are many morphological homologues. It is concluded that, at least in *Schefflera*, some three stages of speciation may be recognized, the last possibly still continuing.

David G. Frodin, 2401 *Pennsylvania Avenue, Philadelphia, PA 19130, USA, Telephone* 215-232-3269.

LECTURE

Patterns of Diversity and Floristic Composition in Neotropical Montane Forests

A. Gentry

Data for plants >2.5 cm dbh in 0.1-ha samples of neotropical montane forests are compared. The available data set includes 32 Andean sites representing 17 Departments in four countries as well as 11 Central American/Mexican sites. Diversity of Andean forests decreases linearly with altitude above 1500 m; up to 1500 m Andean forests are as diverse as lowland tropical forests. Central American montane forests, like their lowland equivalents, are generally less diverse than are similar South American forests. Up to 1500 m Andean forests are floristically similar to lowland Amazonian forests; above 1500 m they are composed of a very different set of predominantly Laurasian families and genera. Different Andean forests at similar elevations are remarkably similar in their floristic composition at the familial and generic level. Lauraceae is the most speciose woody family in virtually all Andean forests between 1500 and 2900 m elevation, followed by Melastomataceae and Rubiaceae. At high altitudes near timberline Compositae and Ericaceae become the most speciose elements of the woody flora. Despite the generality of these patterns the species that compose different Andean forests show little overlap.

Alwyn Gentry, Missouri Botanical Garden, St Louis, MO 63166-0299, USA.

POSTER

Flora, Structure and Composition of a Secondary Fragmented Forest in Antioquia, Colombia.

D. Giraldo-Cañas

Flora, structure, and composition of three remnants at different successional stages of a secondary forest, located in Río Santo Domingo cañyon, along Magdalena Medio, Central Cordillera of Antioquia, was studied. Flora analysis was carried out in 2.5 hectares. For structure analysis, transect technique was used (50 × 2 m); the total area analyzed was 1200 m² determined by Shannon-Weaver accumulative index (H'). Individuals with a dbh greater or equal to one inch were included. Flora analysis revealed 811 species distributed among 176 families out of which 48 are non-vascular cryptogams with 114 species, 17 vascular cryptogams with 69 species, 18 monocotyledon families and 93 dicotyledons with 136 and 492 species, respectively. Structural analysis revealed 514 individuals belonging to 117 species distributed among 49 families. About 76% of the species were represented by less than three individuals suggesting a homogeneous distribution. Peilou's evenness index (J') showed values close to the unit (1), a fact which suggests low variation. In the same way, Simpson's index (D') revealed low dominance, contrary to what is expected for secondary forest. Numeric species richness (s) per transect was found to be highly, but inversely, associated to H' as well as D'. On the other hand, J' was not linearly related to H'. Only five species were found in all arboreal strata and 13 more were found in only three of them. Though their relative abundance in percent was low, it can be inferred that these species will have a place in the structure and composition of the forest. This study also shows that the survival of the rest of the species is questionable. Flora similarity was estimated based on Czekanowski's index, and values were also low; lower than 21% among transects and arboreal strata, and lower than 10% among remnants. This is evidence of great variation in flora composition in the studied area. It can be concluded that the diversity of this forest is a function of the great quantity of herbaceous and shrubby plants of different habits, and not a function of woody species. Ecological indices revealed a high number of species with a very homogenous abundance. Statistical analysis showed that there is not a significant difference among the different ecological indices among remnants. This suggests that the most advanced succesional stages are not necessarily more diverse than the younger ones.

Diego Giraldo-Cañas, University of Antioquia, Medellin, Colombia.

LECTURE

Typology of Flowering Units of the Species of Hypericum from the Eastern Cordillera, Colombia

F. Gonzales and L. E. Mora-Osejo

Typology of inflorescences, based on the Troll and Mora-Osejo models, is used to study the growth form and the flowering units of species of *Hypericum* from the Eastern Cordillera, Colombia. *Hypericum* sect. *Brathys* is most diverse in Colombia; the species are concentrated in the upper edge of high-Andean forests, as well as in the páramos. They are, mostly, shrubs, shrublets or wiry herbs, with monotelic inflorescences. However, a notable variation of the type of branching and the position (incl. polytelec unites) and anthesis of flowers if recognized in Colombian species. A comparison of vegetative and flowering zones to each species is given. Results are accompanied by autecological observations.

Favio Gonzales, Instituto de Ciencias Naturales, Universidad Nacional de Colombia, Bogota, Colombia. Luis Eduardo Mora-Osejo, Jardín Botánico de Bogotá "José Celestino Mutis," Bogotá, Colombia.

LECTURE

Vegetation Changes Along an Andean Mountain Landscape, Tachira State, Venezuela

V. González and M. Pietrangeli

The work was carried out taking as study area, the basin of the river El Valle (9745 ha) in the Tachira State, Venezuela (7°57'–8°05'N; 72°– 72°05'W). The research area represents a typical mountain landscape in which five different kinds of geological formations outcrop. The older one, from the Precambric era, is composed of igneous-methamorphic rocks (granite, gneiss, and schist), while the younger ones, from the Paleozoic and Mesozoic era, are composed by a sequence of sedimentary rocks in which shale, limonite and sandstone, alternate with each other and are interrupted by limestone lens. The mean annual rainfall increases altitudinally and from north to south. The extreme values vary from 500–1200 mm. The mean annual temperature varied from 8.5–24.5°C. The main objective of the study was to characterize the different kinds of vegetation along the mountain landscape and across the altitudinal gradient which varied from 1150–3150 m elevation. The results showed the presence of different kinds of communities along the altitudinal gradient. These varied from cactus scrub along intramontane valleys associated with semiarid environment (rain shadow effect), up to a páramo vegetation, above 3000 m elevation. Between both extremes, we were able to recognize six different kinds of tropical mountain forests. In the present lecture we discuss the possible causes that condition the floristics and physiognomic changes along the altitudinal gradient. These are related to the increase of total rainfall along the altitudinal gradient; at the same time, the air temperature decrease 0.8°C per 100 m altitude. The amount and the percentage of cloudy hours also increase as well as the interception by the vegetation (hidden rainfall). The interaction of these variables decrease the rate of organic matter decomposition with increasing altitude and slow down the nutrient cycling. At the same time, the increase of the wind action along the altitudinal gradient, causes gradual reduction of the structural complexity and floristic composition of the mountain forests. Project supported by CONDES (870-91), Univ. del Zulia; Centro Ecologico de las Tierras Altas, El Cobre, Táchira and FUNDACITE-Táchira.

Valois González, Univ. Central de Venezuela, Fac. de Ciencias, Inst. Zoología Tropical, Apdo. 47058, Caracas, Venezuela. Miguel Pietrangeli, Univ. del Zulia, Fac. de Ciencias, Dpto. de Biología. Apdo. 526, Maracaibo 4011, Zulia, Venezuela.

LECTURE

Diversity of Andean Hepaticae

S. R. Gradstein

The Andes, with an estimated 700–800 hepatic species in 135 genera and 39 families, harbour the richest hepatic flora of the neotropics. The greatest diversification is in the humid montane forests where hepatics outnumber mosses and where large epiphyte masses of hepatics may play an important role in the hydrological balance of the forests. Transect studies in Colombia (ECOANDES project) and Peru (BRYOPTROP project) indicate that the majority of the species have wide ranges throughout the cordilleras. Local floristic differences seem to be due mainly to ecological factors. The rate of endemism increases with altitude and is higher than elsewhere in the neotropics, in spite of the young age of the cordilleras. The vast majority of the Andean hepatics are of tropical origin. They probably arrived from lower altitudes during the upheaval of the cordilleas in the late Tertiary. Temperate species are rather few and occur mainly above 3000 m. They should have arrived after the upheaval of the cordilleras. A few of them seem to have invaded the Andes in recent, Holocenic times due to manmade disturbance of the environment. Although our understanding of the hepatic diversity of the Andes has increased considerably in recent years, knowledge is still hampered by the lack of taxonomic treatments of some major groups, especially the Plagiochilaceae and part of the Lejeuneaceae.

S. Rob Gradstein, Herbarium, University of Utrecht, Heidelberglaan 2, 3584 CS Utrecht, The Netherlands. Telephone 31-30-531816/210470, Fax 31-30-518061.

LECTURE

Relationship Between Northern South American/Southern Central American Vegetation During the Tertiary

A. Graham

A compilation of 636 palynomorphs representative of Cenozoic deposits in northern South America is compared with 358 types from southern Central America. In an age by age comparison the percent similarities are 2.6% for the middle(?) to late Eocene, 10.7% for the lower Miocene, and 8.9% for the middle Pliocene. There are two implications of this new data for the geology and biogeography of the region. One relates to the initial time of substantial interchange of biotic elements that lack effective means of dispersal across the moderate marine barriers that existed between North and South America for most of the Cenozoic. This time has been estimated within a range of about 5–1.8 Ma, based on geologic reconstructions and paleontological evidence. The middle Pliocene Gatun flora of central Panama (*ca.* 4 Ma) is remarkably distinct from floras of similar age in northern South America. An age for closure toward the younger part of the age range is in better accord with the developing paleobotanical evidence. Recent paleofaunal studies place the time for closure at about 2.4 Ma. The other implication relates to the degree to which the isthmian land bridge served as a general migration route between the two biotas during the Tertiary. With regards to lowland elements, *Nypa* is known from the Eocene Laredo Formation of south Texas and is widespread in Tertiary deposits from northern South America. A report from the Oligo–Miocene La Quinta Formation of Chiapas, Mexico, suggests it might have been present and migrated through the region, but a re-evaluation of that report indicates the specimens are not *Nypa* (they may represent some member of the Malvaceae). There are no other fossil records of *Nypa* in Mexico or Central America. With reference to montane elements, the two regions were distinct throughout most of the Cenozoic because altitudes up to only about 1200–1500 m were present until the middle Pliocene, when they increased to about 1700 m (current maximum elevations are 3500–4000 m). This raises a question about the report of *Nothofagus* in the Eocene of SE United States because suitable habitats were absent in most of the intervening region of Mexico and Central America, and there have been no subsequent confirming records. Collectively, the paleobotanical evidence, in accord with recent geologic and other paleontological data, is beginning to suggest that the isthmian land bridge is a recent and ephemeral structure, and has been a more selective avenue of dispersal than implied in the older biogeographic literature, particularly with regards to montane elements.

Alan Graham, Department of Biological Sciences, Kent State University, Kent, OH 44242, USA. Telephone 216-672-7888, Fax 216-672-3243, Email: agraham@kentvm or agraham@kentvm.bitnet.

LECTURE

An Overview of Neotropical Montane Mushrooms (Fungi, Agaricales)

R. E. Halling

Agarics of the Neotropical montane forests show several different patterns of origin and distribution. These patterns are often directly related to the types of habitats in which the agarics occur and are influenced by geological history, phanerogamic vegetation, and climate. Those agarics that form ectomycorrhizae appear to be narrowly restricted to forests of *Quercus*, *Alnus*, and *Salix* and have their origins and/or affinities with north temperate taxa. On the other hand, the saprotrophic (litter decomposing) agarics are widely distributed in anectotrophic forests as well as in ectotrophic forests. Thus, the litter decomposers will have a broader range of distributional groupings that may be categorized as Amphiatlantic, Neotropical, Pantropical, and Cordilleran, and can also include local Gondwanan endemics.

Roy E. Halling, Institute of Systematic Botany, New York Botanical Garden, Bronx, NY 10458-5126, USA. Telephone 718-220-8613, Fax 718-562-6780, Email: rhalling@nybg.org .

LECTURE

Environmental Impact of Coca and Cocaine Production in the Chapare Region of Bolivia

R. Henkel

The purpose of this presentation is to describe and analyze the impact of coca and cocaine production on the environment of the Chapare, a tropical lowland rain forest in Bolivia. An analysis of data from Bolivian government agencies on coca and cocaine production and data collected from farmers and cocaine producers indicate that the "boom in coca" in the Chapare Region during the 1980's produced few negative impacts on the environment. Most of the environmental impacts were of a positive nature resulting from farmers reducing their cultivation of other crops and concentrating on production of coca. Vast areas in the older settled areas of the Chapare reverted to second growth forest as farmers abandoned the cultivation of rice, bananas, citrus and other crops to concentrate on the production of coca. Also very little undisturbed forest land was cleared for coca in that most new coca fields where developed in the older settled regions already under cultivation. The increase in second growth forest resulted in improved habitat for wildlife and birds. It also greatly reduced erosion and river sedimentation which had a positive effect on aquatic species. The reduction in sediment load and rate of sediment deposition in rivers also reduced the potential for flood damage. Virtually the only negative impact on the environment resulted from the dumping of sulphuric acid, sodium carbonate, kerosene, gasoline and other chemicals used in extracting cocaine from the coca leaves. Due to its high market value and the small amount of land required to support a family, coca cultivation is much less damaging to the environment than crops formerly cultivated in the Chapare Region or proposed for cultivation under the U.S.-Bolivian crop substitution program.

Ray Henkel, Department of Geography, Arizona State University, Tempe, Arizona 85287-0104, USA. Telephone 602-965-7533, Fax 602-965-8313.

Evolutionary History of North Andean Montane Forests and Quarternary Climatic Change

H. Hooghiemstra

The pollen records of two sediment cores 357 m and 586 m deep, from the high plain of Bogotá (Eastern Cordillera, Colombia, 2550 m elevation), show the history of the montane forests and alpine grassland vegetation of this area during the last three million years. Several phases in the evolutionary history can be recognized. Late Pliocene and Quaternary climatic change and the Panamanian land bridge, which came into existence around 5–4 Ma, had a strong influence on the development of these montane ecosystems. The period 3.2–2.7 Ma shows a warm climate most of the time (upper forest line oscillated from 2800 to 3600 m). The basin had just started to accumulate sediments, after the final upheaval of the Eastern Cordillera. The sub-Andean (lower montane) forest belt reached some 500 m lower than today. With increasing altitude *Podocarpus*-rich forest, *Hedyosmum-Weinmannia* forest, *Vallea-Miconia* forest, and *Hypericum-Myrica* dwarf forest near the upper forest line, were the major forest types of the Andean (upper montane) forest belt. These forests were more open than at present-day and the heliophytic element *Borreria*, originally a savanna element, was abundant. Arboreal pioneers (*Dodonaea*, *Eugenia*) and *Symplocos* and *Ilex* contributed to zonal forests and also constituted azonal forests of a rapidly changing composition. *Alnus* and *Quercus* were absent. In the period 2.7–2.2 Ma *Hedyosmum* dominated in the Andean forest belt for the first time in the record. *Miconia* dominated the *Vallea-Miconia* forests, in which *Ilex*, *Rapanea*, and *Daphnopsis* were associated. For the first time in the record, páramo vegetation became widespread in the area. The composition of the páramo flora increased gradually in diversity and taxa such as *Valeriana*, Caryophyllaceae, *Aragoa*, and *Plantago* were common. The period 2.2–1.42 Ma shows for the first time a persistent cold climate and the upper forest line oscillated from 1900 to 2500 m, most of the time. Andean forests became denser in structure, *Daphnopsis* almost disappeared *Juglans* appeared for the first time regularly, *Styloceras* became more common, and *Polylepis* dwarf forest started to increase near the upper forest line. *Plantago rigida* cushion bogs became abundant in the basin. The period 1.42–1.0 Ma shows a cold climate most of the time. *Weinmannia* was almost absent in the Andean forest belt. *Hedyosmum* forest with *Eugenia* and *Rapanea* constituted a precursor of the modern *Weinmannia* forest. *Vallea* contributed for the first time in the record as much as *Miconia* to the Andean forest. *Myrica* and *Polylepis* were abundant in timberline forests. The upper limit of sub-Andean forest belt reached higher altitudes (but should reach modern elevations at only 120 ka). The lake was shallow with abundant marsh vegetation, but possibly after a tectonic event at c. 1 Ma, the lake became deeper. During the period 1.0–0.56 Ma the first major glacial-interglacial cycles, comparable to the Late Quaternary, occurred. A 100 ky frequency became dominant in the altitudinally shifting upper forest line.

Alnus, a northern hemisphere genus, immigrated along the Panamanian Isthmus into the area. When the high plain was situated in the Andean forest belt, during interglacial phases, *Alnus* swamp forest vegetation came into existence on the wet flats around the lake and replaced vegetation of *Myrica* thickets. Most of the time, parts of the lake were shallow with abundant marsh vegetation, alternating with periods of high water levels during cold intervals. During the period 0.56–0.34 Ma the belt with *Polylepis* dwarf forest and *Hypericum* scrub vegetation (subpáramo belt) became more important during a cold climate most of the time. *Weinmannia-Hedyosmum* forest with *Rapanea, Vallea-Miconia* forest, but also *Myrica* and *Alnus* contributed to the Andean forest belt. The period 0.34–0.186 Ma shows a warm climate most of the time. *Quercus,* also a characteristic northern hemisphere genus, immigrated into the area of Bogotá about 0.34 Ma. During southward migration through Andean Colombia, greatest horizontal displacement occurred probably at lower elevation, in the upper sub-Andean to lower Andean forest belt, and subsequently *Quercus* expanded locally and regionally to higher altitudes. At first local patches of oak forest occurred in the area of the high plain and *Quercus* contributed since only 0.2 Ma with substantial zonal forests in the Andean forest belt near Bogotá. *Acalypha* and *Alchornea* reached in the *Quercus* forests higher elevations and the sub-Andean forest belt reached modern altitudinal distribution since the last interglacial (c. 125 a). *Weinmannia* and *Miconia* became important elements of the (upper) Andean forest. The period 0.186 Ma–recent shows for the first time abundant presence of *Quercus* forest, leading to the modern *Saurauia-Quercus* forest type. The natural composition of the Andean forest belt had changed dramatically for the last time. The Late Pliocene–Quaternary pollen record of Colombia evidences a continuous change in the composition of montane forests. In this process, evolutionary adaptation, migration and climatic change are the most important factors. This record of continuous change forms a framework in which modern aspects of biodiversity can be placed.

H. Hooghiemstra, Hugo de Vries Laboratory, Dept. of Palynology and Paleo/Actuo-Ecology, University of Amsterdam, Kruislaan 318, 1098 SM Amsterdam, The Netherlands. Fax 31-20-5257715.

LECTURE

A General Overview of Montane Forest Types in the Venezuelan Guayana

O. Huber

A new vegetation map of Venezuelan Guayana (*ca.* 450,000 km^2), gives a detailed classification of forest types and associated altitudinal zonation. The region includes three main physiographic zones: 1) macrothermic lowlands, *ca.* 0–500 m; 2) submesothermic uplands, *ca.* 500–1300/1500 m; 3) meso- to submicrothermic highlands, *ca.* 1500–2500/3000 m. The lowlands consist of large plains and peneplains, alternating with low hills; the uplands comprise higher hills, intermediate mountain slopes, and high plains, whereas the highlands consist mainly of the upper slopes and mostly flat-topped summits of the characteristic Guayanan table mountains ("tepuis") of the Roraima sandstone formation, and some granitic high mountains. Forest cover is most extensive in the lowlands and uplands, while in the highlands other, non-forest vegetation types dominate. Montane forests of the Venezuelan Guayana show a wide range of little known floristic, physiognomical, and ecological variation. Some characteristic features, however, can be recognized and classified according to the following general altitudinal zonation scheme: 1. *Basimontane forests*, extending roughly between 100–500 m. These usually dense, medium-sized forests represent transitional communities from lowland forests to true montane forests; they range from semi-deciduous to evergreen; their main families are Leguminosae, Moraceae, Annonaceae, Lecythidaceae, Burseraceae, Bombacaceae, and palms. 2. *Submontane* (or *lower montane*) *forests*, represent the bulk of the Guyanan montane forests, extending widely on lower slopes and uplands between *ca.* 500–1200 m. They are often tall to very tall (up to 60 m), relatively dense, and mostly evergreen. Dominant families are Sapotaceae, Melastomataceae, Rubiaceae, Leguminosae, Vochysiaceae, Lauraceae, and palms. 3. *Upper montane* (*tepui) forests*, are evergreen, low to medium-sized dense forests covering the upper mountain slopes at 1200–1800 m. They vary greatly, both in physiognomy and in floristic composition, from one mountain system to another, depending on the type of substrate, exposure, *etc.* Characteristic families are Ochnaceae, Podocarpaceae, Magnoliaceae, Sapotaceae, Steraceae. 4. *High montane (tepui) forests*, occur scattered on the tepui summits and along the upper rocky slopes, mostly at 1800–2500 m. These low, evergreen and typically sclerophyllous forests are dominated by Theaceae, Clusiaceae, Ochnaceae, Asteraceae, and Araliaceae. Most montane forests of the Venezuelan Guyana are protected in extensive national parks and natural monuments. In some areas (Parima uplands, northern hills), deforestation is severe, leading to establishment of secondary shrub lands and grasslands.

Otto Huber, GTZ/CAIAH. Apartado 80405, Caracas 1080-A, Venezuela. Telephone and Fax 58-2-772528.

LECTURE

Epiphyte Composition at Monteverde, Costa Rica

S. Ingram and N. Nadkarni

The purpose of our study was to document the floristic composition of the vascular and non-vascular epiphyte communities in a lower montane rain forest, Costa Rica. The four-hectare study site, located within the Monteverde Cloud Forest Reserve "leeward cloud forest" research area, includes many identified tree species, and 44 canopy-level trees rigged for climbing with Jumar ascenders. We used opportunistic collecting techniques (collec-ting specimens from lower tree trunks, fallen branches, and treefalls) and single-rope climbing techniques to collect from canopies of 38 standing trees. Information on host species, branch angle and circumference, and substrate type and depth were recorded for each specimen. We also noted the within-tree distribution and substrate characteristics of six common "target" species. Target species were ecologically and taxonomically diverse, yet readily identifiable in sterile condition. Plant and data collections were made during four one-month trips at different times of year between February 1992 and April 1993. Approximately 300 species of vascular epiphytes from 23 angiosperm families and 18 pteridophyte families, and 60 non-vascular species (bryophytes, foliose and fruticose lichens) were found within the study area. The angiosperm families Araceae, Bromeliaceae, Ericaceae, Gesneriaceae, Orchidaceae, and Piperaceae accounted for 80 percent of angiosperm species, and orchids made up nearly half of the angiosperm flora. Of the epiphyte species 60% were collected from fallen trees and branches, 25% from lower tree trunks, and 15% from within standing tree canopies only. The target species *Pleurothallis ruscifolia* and *Disterigma humboldtii* were found primarily in thick humus mats in the upper canopy. The target species *Anthurium scandens*, *Guzmania angustifolia*, and *Columnea microcalyx* were found most often at lower and mid-canopy levels, and *Campyloneurum sphenodes* was most common on lower tree trunks. By collecting basic ecological information with plant specimen collections and with target species observations, we were able to discern distributional patterns of a few epiphyte species. Collecting from standing tree canopies at different seasons over two years enabled us to produce a preliminary epiphyte species checklist nearly as complete as a long-term general plant collection would provide.

Stephen W. Ingram, Marie Selby Botanical Gardens, 811 South Palm Ave., Sarasota, FL 34236, USA. Telephone 813-955-7553, Fax 813-951-1474. Nalini M. Nadkarni, MS Lab II, The Evergreen State College, Olympia, WA 98505, USA.

POSTER

Affinity Among Mountain Ranges in Megamexico: A Phytogeographical Scenario

G. Islebe and A. Velázquez

A large number of recent publications have pointed out the floristic resemblance among mountain ranges of Meso-America. Most authors based their findings on presence/absence data of plant genera. Because of the floristic similarity, most mountain regions in Mexico and Guatemala are considered as a single unity, currently called "Megamexico 2" (Rzedowski, 1991 in: Act. Bot. Mex. 14:3–21). Until now, no systematic research has been done to measure this floristic affinity. Recent surveys, carried out in mountain ranges of Mexico (Sierra Chichinautzin and Sierra Nevada) and Guatemala (Cadena Volcánica and Sierra de los Cuchumatanes), provide data to document the floristic relationship between these regions. In the present investigation it is hypothesized that no substantial floristic differences are to be expected, and if so, nor major climatic and ecological distinction too. To test this hypothesis, we followed a deductive approach and statistical methods. A cross-section in every mountain range was made in order to sketch and to compare the present altitudinal distribution of the different vegetation types. Data from 40 representative relevés (*sensu* Zürich-Montpellier school), which characterize every zonal high mountain vegetation type, were used to measure the statistical (dis)similarity among mountain ranges. The analysis included the performance of detrended correspondence analysis and the computation of the Shannon diversity index. The output provide evidence to accept the hypothesis mentioned above. In other words, the preliminary results indicate that no major generic floristic differences among these mountain ranges are present. Further observations on fauna elements shared by these areas, and man-made activities, are given. This research documents the highly threatened status of these insular and last remaining natural areas. Finally, it is recommended to strengthen the conservation struggle of these ecosystems by implementing adequate management and protection plans.

Gerald Islebe, Hugo de Vries Laboratory, University of Amsterdam. Kruislaan 318, 1098 SM, Amsterdam, The Netherlands. Telephone 31-20-525-7193, Fax 31-20-525-7715. Alejandro Velázquez, Laboratorio de Biogeografía, Facultad de Ciencias, UNAM. Crúz Azúl nr. 14, C. P. 14370 México D. F., México. Telephone 525 6719906, Fax 525 548 8186. Email: Alexvela@sara.nl.

POSTER

Plant Succession in a Subtropical Montane Rain Forest of Northwestern Argentina

C. A. Iudica and T. M. Schlichter

The purpose of this study was to determine the successional trends of primary and secondary physiognomies of a montane rain forest. Based on the structure of these two communities we postulated ways in which succession affects the pattern of natural disturbance regeneration. Recurrent disturbances in a montane rain forest are frequently the beginning of a secondary succession. This succession could evolve into a community similar to the original (convergence) or stabilize in a different condition (divergence). The data for this paper was obtained during October 1987 in the Montane Forest District (Distrito de las Selvas Montanas, Cabrera 1976) of the phytogeographic Yungas Province (Cabrera 1976) in the Calilegua National Park. The Park is located at 23°30'S, 64°45'W within Ledesma County, Jujuy Province, Argentina. In sixteen 0.1 ha plots we compared structural elements (basal area, crown volume, number of strata, diameter classes, and complexity and diversity indexes) and floristic composition of primary and secondary forests. One of the analyzed stands corresponds to a mature community, representative of the typical physiognomy of primary forest. The other stand corresponds to a mature secondary forest found on the scar left by a natural avalanche. According to different responses of the species to avalanches we defined three classes of species (harmed, favoured, and indifferent species). We also characterized structure and floristic composition of both primary and secondary forests. Analysis of the effects of debris avalanches on the secondary community showed that there is a clear successional trend from the secondary to the primary community. The debris avalanches released resources that would be used by the secondary community and would determine its future successional path. The explained phenomenon is recurrent and decisive in the dynamics of montane forest regeneration.

Carlos Alberto Iudica and Tomas Mario Schlichter, Tropical Conservation and Development Program, Center for Latin American Studies, 319 Grinter Hall, University of Florida, Gainesville, FL 32611, USA. Telephone 904-392-0375/6548, Fax 904-392-0085, Email: casaiud@pine.circa.ufl.edu.

LECTURE

Useful Forest Trees of the Andean Slopes in Ecuador

J. L. Jaramillo

The eastern and western slopes of the Ecuadorian Andes, between 1600 m and 3000 m elevation, harbour a number of species of forest trees that are of interest locally and regionally for the production of timber, firewood, and other useful products. These species occur mostly on steep terrain with fragile hydromorphic soils. During the past several decades, vast areas of the Andean forests have been cleared by colonists who, after selling off the timber, have converted the land to pastures and crops; natural vegetation has largely been destroyed. A more rational use of these fragile lands would entail the production of useful tree species. The colonists already possess some knowledge of forest management that could form the basis for sustainable silviculture in the region. Floristic studies of the Andean forest tree species during a 15-year period, as well as interviews with local inhabitants and personal observations, have revealed a total of about 50 species of fast-growing, easily propagated native species of trees that could be incorporated into silvicultural systems for the production of timber and fuelwood. Some of the most promising species, which can produce marketable volumes of timber on 10–15 year rotations, include *Alnus acuminata* (Betulaceae), *Croton dacryodes*, *C. suribus* and *Sapium verum* (Euphorbiaceae), *Guarea kunthiana* and *Carapa guianensis* (Meliaceae), *Freziera reticulata* (Theaceae), *Bombacopsis squamigera* (Bombacaceae), *Trema micrantha* (Ulmaceae), and *Morus* sp. (Moraceae). Further studies are required on the reproductive biology, growth characteristics, and management of these and other species. Such studies would allow the development of silvicultural systems to restore forest cover on the deforested Andean slopes and provide economic benefits for the local inhabitants.

Jaime L. Jaramillo A., Herbario QCA, Pontificia Universidad Católica del Ecuador, Casilla 17-01-2184, Quito, Ecuador. Telephone 593-2-529270, ext. 279, Fax 593-2-567117.

LECTURE

Floristic Analyses of High-montane Ecuador

P. M. Jørgensen, C. Ulloa Ulloa, R. Valencia and J. E. Madsen

The high-montane Ecuador (>2400 m) covers 45,000 km² and a range of climate, geology, and soil conditions. Using general collecting, quantitative inventories, herbarium inventories, and the literature we have compiled information on diversity patterns from this area. It contains 4800 species, 1119 genera, and 200 families of vascular plants; we estimate the total flora of Ecuador at 29,000 species or about 45% higher than previous estimates. The lowest zone (2400–3000 m, 17,000 km²) included 3411 species, or 300 species more than was encountered in 70,000 km² of lowland Ecuadorian Amazonas. The diversity of trees in four one-ha plots is, however, only 10–30% of what is found in the lowland Amazon. The inventory included 1565 woody species in 93 families. Of 292 woody genera there were 18% endemic to neotropical Andes, 44% neotropical, 12% pantropical, 6.7% American-Asiatic, 1% American-African, 17% temperate, and 2% cosmopolitan. All life-forms increased in diversity with decreasing elevation. Herbs, shrubs, and epiphytes are less diverse in the lowlands, while the diversity of trees and other life forms (lianas, hemiepiphytes, saprophytes, etc.) increase in numbers. Relative importance of different life forms in the different zones was also calculated and revealed that the zone at 3400–4000 m, often included in treeless páramo zone, has a relative composition similar to the lower forest zones. Distribution maps of species showed differences in composition between north and south Ecuador, this is explained by three factors: migration barrier in and around the Girón-Paute valley both during the ice-ages and at present, differences in the geological deposits, and lower precipitation and a pronounced dry season in the south-western part of the country. It can be concluded that: the montane flora of Ecuador is extremely rich, even richer that the Amazon lowland; more than 60% of the woody genera in the study area are unique for the neotropical area; the maximum diversity of herbs, shrubs, and epiphytes is found at 600–3000 m elevation, while maximum diversity of trees is found in the lowlands; the differences in floristic composition between north and south Ecuador can be explained by the three factors: precipitation, geology, and migration barriers.

Peter Møller Jørgensen and Carmen Ulloa Ulloa, Missouri Botanical Garden, P.O.Box 299, St. Louis, MO 63166-0299, USA. Telephone 1-314-577-5100, Fax 1-314-577-9596, Email: jorgense@mobot.org. Renato Valencia, Herbario QCA, Departamento de Ciencias Biológicas, Pontificia Universidad Católica del Ecuador, Apartado 17-01-2184, Quito, Ecuador. Telephone 593-2-529-260 ext. 279, Fax 593-2-567-117. Jens E. Madsen, Institut des Sciences de l'Environnement, Universiti Cheik Anta Diop, Dakar, Senegal. Telephone 221-242302, Fax 221-242302.

LECTURE

Diversity along a Successional Gradient in a Neotropical Montane Rain forest, Costa Rica

M. Kappelle, P. A. F. Kennis and R. A. J. de Vries

Recovery following clearing of a neotropical montane rain forest has been studied in the Costa Rican Cordillera de Talamanca. Special attention has been paid to species diversity of terrestrial vascular plants. Different successional forest phases (pioneer, early successional, mid-successional, and mature) have been distinguished. Results show an increase in species diversity during succession, being largest in the mid-successional phase. Families such as Asteraceae, Campanulaceae, Rosaceae, Rubiaceae Scrophulariaceae, and Solanaceae dominate the early secondary phases, while Araliaceae, Melastomataceae, Myrsinaceae, and Onagraceae become prominent in later phases. The mature forest phase is characterized by the fagaceous genus *Quercus*, accompanied by families such as Clusiaceae, Ericaceae, Lauraceae, and Symplocaceae, which already appear in later successional phases. In conclusion, secondary forests in montane Costa Rica appear to be richer than primary (mature) forest. This seems to hold true for the terrestrial vascular flora as such, thus not taking into account the rich epiphytic vascular flora, that inhabits the mature montane forest canopy.

Maarten Kappelle, Peer Kennis and Rob de Vries, Hugo de Vries Laboratory, University of Amsterdam, Kruislaan 318, 1098 SM Amsterdam, The Netherlands. Telephone 31-20-5257830 or 31-20-6720500, Fax 31-20-5257715.

LECTURE

Botanical Inventory and Conservation

S. Keel

In many cases, the identification of priority conservation sites and the planning and management of protected areas depend on the availability of botanical data. Developing inventory techniques to obtain information applicable to biodiversity conservation has been a major goal of The Nature Conservancy (TNC) for many years. Principally, the inventory methods currently used by TNC fall into two categories: database networking and rapid ecological assessment (REA). TNC's botanical database is an inventory of plant diversity at species and infraspecific level. The information is derived mostly from published taxonomic monographs and regional floras. A different approach informs REA work: its botanical inventory begins on the landscape or community level and then seeks out information on target species. REA is used in cases where little or no botanical information is available, serving as a tool for conservationist to remedy information gaps. These two inventory approaches have been used by many Conservation Data Centers (CDC) in the Neotropics. Two surveys of montane systems carried out by the CDCs of Guatemala and Ecuador will be used by the author to show the usefulness of these inventory methods to plant conservation.

Shirley H. Keel, Latin America Science Program, The Nature Conservancy, 1815 North Lynn Street, Arlington, Virginia 22209, USA. Telephone 703-841-2714, Fax 703-841-2722.

LECTURE

Present and Potential Distribution of Polylepis Woodlands in the Bolivian Highlands

M. Kessler

Patchy distribution of *Polylepis* woodlands in the Andes has been explained by restriction to microclimatically favoured biotopes. Recently, human over-exploitation of Andean ecosystems has been said to be responsible for the limited distribution of these woodlands. To distinguish between these factors, I studied which ecological factors naturally limit the distribution of *Polylepis*. Some correlation was found between occurrence of woodlands and microhabitats such as rocky slopes, boulder screes, and stream margins. But often woodlands were found under quite different conditions. Within the zonal distribution of *Polylepis* (about 3000–4000 m in the E Highlands) any mountain slope, regardless of inclination, exposition, soil type, geological underground, or rockiness may support woodlands; limiting factors were salty or wet soils in valley bottoms. No woodlands were found on the Altiplano, except near Lake Titicaca. In the W Cordillera woodlands were mostly restricted to volcanic slopes at 4200–4800(–5000+) m; here the lower limit was determined by cold air accumulation at night and salty soils. At the limits of its ecological amplitude, *Polylepis* was restricted to extrazonal locations; here preference of particular expositions (*e.g.*, well-insolated N-exposed slopes in the cold W Cordillera) or rocky slopes (at the driest localities) was evident. No correlation was found between occurrence of other edaphically azonal biotopes and the proximity to the distributional limits of *Polylepis*, suggesting that there is no microclimatic favouring of these localities. Close correlation was found between human disturbance and age structure of *Polylepis* woodlands. Burning and grazing are the most destructive factors, restricting woodlands to sheltered locations (stream margins, rocky slopes). In the inhospitable W Cordillera the area covered by *Polylepis* woodlands (5000 km²) is probably close to the potential distribution; the densely settled Altiplano is considered to he potentially free of *Polylepis*. In the E Highlands *Polylepis* woodlands could potentially cover 50,000 km², but cover only 600 km² (1.2%) of the potential extension. *Polylepis* woodlands could cover about 55,000 km² (20.5%) of the Bolivian Highlands (265,000 km²). The remaining area is either too high (11.0%) or too low (18.1%) for *Polylepis*, encompasses the Altiplano (44.5%), or is otherwise unsuitable (wet or salty soils, arid regions, *etc.*). Natural factors (fires, browsing by native animals), as well as early human disturbance might have prevented *Polylepis* from reaching its potential distribution after the last glacial period.

Michael Kessler, *Schiefer Weg 5, 3400 Göttingen, Germany. Telephone 0551-704914.*

LECTURE

The Effect of Cutting and Grazing on Upper-Andean Forestline Vegetation

K. Kok

This study describes the vegetation of the Colombian forestline and analyzes impact of human influence on its structure and composition. The study area is located in the Central Cordillera in Colombia at 3400–3800 m in the high Andean forest belt. The vegetation is a two-layered forest dominated by *Weinmannia mariquitae* and *Miconia* spp. Grazing was extensive, and its intensity positively correlated to impact of the cutting system practised. Cutting intensities were divided in four classes according to a structure classification of the samples. As occasional removal of trees (selective cutting) or shrubs ('socola') were not believed to influence floristic composition, samples of them were taken to represent natural forest. Totally cleared sites had a regeneration time of five(+) years. Using TWINSPAN (a cluster program), 56 samples covering a range of different cutting systems and grazing intensities were analyzed. The analysis included 74 families, 189 genera and 298 vascular plants; most important families were Asteraceae, Poaceae, and Ericaceae. Important genera were: *Miconia, Mikania,* and *Lachemilla*. Nine vegetation types were distinguished, four of them representing natural forest; five representing regenerating cut forest. Altitude was the main factor explaining the variation between the types. The fourth type of forest represents remnants of *Polylep-is sericea* forest, in which both *W. mariquitae* and *Miconia* spp. are absent. The five disturbed types were all related to one of the first three natural forest types. Species composition is altered considerably; main floristic differences are found in the presence of approximately 20 pasture species and the absence of most tree species, even in the herb and ground layer. Two distinct situations are two types of totally cut high upper-Andean forest, the first with an abundant regrowth of woody species including the important tree species, and the second with an invasion of various elements from the shrub páramo flora. To establish the ranges of the different types, cover of the main woody species was depicted against altitude for natural forest types and for regrowth after cutting. A combination of altitudinal position and floristic composition provided a model with the possible transitions. Concluding, an altitudinal range of 400 m is broad enough to include considerable differences in floristic composition. Cutting at high altitudes leads to a lowering of the forestline. The effect of cutting at lower altitudes is limited to the temporal removal of trees after which regeneration of the original forest seems possible, especially at intermediate altitudes and when grazing intensities are low. Higher grazing intensities generally lead to the creation of pasture lands independent of altitude.

Kasper Kok, Fazantstraat 171, 7523 DP Enschede, The Netherlands. Telephone 053-356710 (home), 053-874444 (work).

LECTURE

Conservation Policies and Montane Forests in Peru

B. León

During the last 25 years, awareness of and interest in the environment has radically increased in Peru. As a result, today the state protects through national parks, sanctuaries, national reserves, protected forests, reserved zones, and national forests, about 10% of the total area of the country. These good news, however, does not imply that the montane zone and especially areas with forests are adequately represented. Montane areas in Peru are heterogeneous; mountain ranges differ in geological history and features, climatic controls, biological components, and history of human impact. Paradoxically, montane forests account for a small percentage of the total protected area. Although several social actors are important in conserving and protecting montane zones, I focus on the role played by two of them: researchers and policy makers. Both social actors interact at two scales: inner (country, regional and local) and outward. The features of and relationships within and between these two actors have been changing rapidly. Due to the present social, political and economic conditions in the country the goals of these actors are separating, affecting negatively the potential for conservation of montane areas. The future of the conservation of montane forests in Peru requires new strategies at the inner scale; some of these are similar to those needed in other Latin American countries.

Blanca León, Museo de Historia Natural, Av. Arenales 1256, Apartado 14-0434, Lima 14, Peru.

LECTURE

Patterns of Distribution of the Fern Genus Campyloneurum (Polypodiaceae) in the Andes

B. León

The American genus *Campyloneurum* is predominantly tropical. This large genus (47 spp.) occurs in different kinds of habitats, mostly in forests of montane areas at 1000–4500 m elevation. The tropical Andes constitute the most diverse region, containing the highest number of species and of endemics. The geological history of the Andean region is fundamental for understanding the biogeography and speciation of this genus. The present-day distributions of *Campyloneurum* species and species-groups were analyzed within three geographic subdivisions that differ in geology and history of mountain building. Besides these latitudinal patterns, elevational ranges also differ among species. For those species with less than 1000 m elevatio-nal ranges, their latitudinal ranges appear to be more restricted.

Blanca León, Museo de Historia Natural, Av. Arenales 1256, Apartado 14-0434, Lima 14, Peru.

LECTURE

The BIOMA Mountain Program: A Pilot Framework for Action, Research, and Conservation in the Venezuelan Andes

Y. Lesenfants

Communication and multidisciplinary interaction between scientist conducting theoretical investigations, conservationists and development workers attempting to implement on-site plans of action, and local populations is lacking. This problem is severe in the Andes, causing waste of scarce resources and delay in the achievement of essential changes towards sustainable development. The BIOMA Mountain Program by the Venezuelan Foundation for the Conservation of Biological Diversity (BIOMA) is a pilot framework for development programs in the Venezuelan Andes area addressing these problems. The design of the program is based on the concept of action-research which seeks integration of all scientific concepts and practices which fall within the realm of conservation, giving special weight to interaction with local populations. The program is located in one of the most unique environments on the planet: the Andean páramo, in the microbasin of La Toma in the Sierra de la Culata, and includes Andean and high Andean ecological zones (3000 –4700 m). The low part of the basin is occupied by the agricultural community of La Toma. The program has five sub-programs: Community Support, Investigation, Ecotourism, Environmental Education, and Diffusion. To develop these sub-programs the program has physical support structures, including the Permanent Center of Community Attention located in the heart of the La Toma (3200 m), which serves as nucleus of community outreach activities. In the high part of the basin (>3700 m), where BIOMA established in 1987, and has since administered, a private ecological reserve of approximately 12,000 hectares, are the Ecological Research Station and the Visitors Center. Unique natural environments and singular landscapes, endangered endemic species, agricultural problems such as extensive grazing in important yet fragile ecosystems and advance of the agricultural border, lack of infrastructures and services enabling multidisciplinary research in support of environmental planning, lack of opportunities for free exchange of ideas between investigators, conservationists, and campesinos. These are some elements of a complex conservation problem in natural ecosystems and tropical mountain agroecosystems. With this program BIOMA hopes to confront the problems of páramo environments, from a multidisciplinary perspective, where active participation of the rural communities and the contribution of scientific practices are indispensable requirements.

Yves Lesenfants, Fundación Bioma, Av.2 con Calle 41, Urbanización El Encanto, Quinta Irma, Planta Alta, P. Box 676, 5101 Mérida, Estado Mérida, Venezuela. Telephone 58-74-638633, Fax 58-74-638633.

POSTER

"Musgolandia" — *The Natural History of Bolivian Mosses. A Preview*

M. Lewis

The work entitled "La Historia Natural de los Musgos de Bolivia" which is now in page-proof, is the first general text in Spanish that treats the mosses of the Andean countries. The purpose of the book is to provide students and biologists with the necessary tool to identify the majority of mosses found in the Andean countries. The books has five introductory chapters treating such aspects as the most commonly used characters for identifying mosses. The following sections include six florulas from typical Andean environments, with descriptions and illustrations of 125 species. The last sections include a key to the Bolivian moss genera, an English–Spanish glossary, a systematic list with reference to the publications dealing the different genera, a catalogue of species and synonyms for Bolivia, and a broad bibliography. The book has more than 160 plates, illustrated by the author. The history is written, in part, informally with many anecdotes, travel diaries and taxonomic comments.

Marko Lewis, Herbario Nacional de Bolivia, Correo Central Cajon Postal 10077, La Paz, Bolivia. Telephone 591-2-792582, Fax 591-2-359-491.

POSTER

The Yungas Forests of Choquercamiri, Bolivia: Legends, Exploration, Botany and Conservation

M. Lewis, E. Garcia, V. Kuno and H. Davidson

Botanical explorations of the Yungas of Choquercamiri have led to the discovery of legendary lost cities. This 7200 km² area of puna and forest is located 90 km SE of La Paz. Choquercamiri is exceedingly rugged, with an altitudinal range of 1000–6000 m. It is highly diverse in Andean forest types and species, especially cryptogams, orchids, and ferns. It is also one of the few natural habitats in which the spectacled bear *(Tremarctos ornatus)* is common. During the past three years we have worked towards the creation of a national park in the area. The creation of a park would preserve the biological and historical value of Choquercamiri and improve the socio-economic situation in the surrounding area. We have proceeded on three levels: 1) Botanical and archeological inventories directed towards identification of the vegetation types and ruins most important for conservation; 2) organisation and involvement of campesino communities and local civic organisations in the planification and implementation of the proposed protected area; and 3) synthesis of scientific data and local planning into a workable conservation plan and model decree. A graphic discussion of these topics will be presented. The present phase of the project includes the description of forest structure and quantification of biodiversity in the region. The method is designed to include cryptogams, epiphytes, vines and herbs. In a *Polylepis besseri—Vallea stipularis* dominated woodland, for example, we discovered that over 50% of the species in a 200 m² plot were cryptogams. Phytogeographic affinities of the moss flora of different forest types in Choquercamiri suggest that neighbouring forest types have different geographical origins. A "kitchen table" discussion of these phenomena will also be presented.

Marko Lewis, Emilia Garcia, Virgilio Kuno and Hilary Davidson, Herbario Nacional de Bolivia, Correo Central Cajon Postal 10077, La Paz, Bolivia. Telephone 591-2-792582, Fax 591-2-359491.

LECTURE

Cloud Forest of "Avila" National Park in the Central Coastal Range of Northern Venezuela

W. Meier

The main types of cloud forests on the southern slopes (1700–2250 m) of "Avila" National Park in N Venezuela above Caracas are: 1) *Clusia multiflora* forests (2000–2250 m); this 1–layered, low (5–15 m tall) forest forms the uppermost forest type limiting the sub-páramo scrub; its main tree species are *Clusia multiflora* (Clusiaceae), *Roupala pseudocordata* (Proteaceae), *Befaria glauca* (Ericaceae), *Gaiadendron tagua* (Loranthaceae), and *Didymopanax glabratus* (Araliaceae). Based on the understory, two subtypes of this forest can be recognized: the interior is dominated by bamboos (*Chusquea* sp. and *Aulonemia subpectinata)*, another subtype has a dense herbaceous layer of *Elaphoglossum* spp. (Pteridophyte), terrestrial bromeliads (*e.g.*, *Guzmania ventricosa)*, and terrestrial orchids (*e.g.*, *Oncidium* sp.) predominates. 2) *Micropholis crotonoides* forests (1800–2000 m); this type is 15–25 m tall and restricted to the exposed upper slopes and ridges. Common species are *M. crotonoides* (Sapotaceae) *Podocarpus salicifolius* (Podocarpaceae), *Sloanea* sp. (Elaeocarpaceae), *Eschweilera tenax* (Lecythidaceae), and *Protium tovarense* (Burseraceae); dense stands of the bamboo *Arthrostylidium venezuelae* are locally dominant. 3) *Myrcianthes karsteniana* forests (1700–1950 m). This tall (20–40 m high) forest is widespread in depressions on the middle slopes of the Avila mountain; most frequent tree species are *M. karsteniana* (Myrtaceae), *Guarea kunthiana* and *Cedrela montana* (Meliaceae), *Pseudolmedia rigida* and *Ficus* spp. (Moraceae), and *Ceroxylon klopstockia* (Arecaceae). A characteristic feature of this forest type is the local dominance of a giant bamboo, *Arthrostylidium* cf. *longiflorum*, reaching up to 10 m, and of many species of Solanaceae in the understory. 4) *Croton–Montanoa* pioneer forest; a gap-filling forest type scattered in the *Myrcianthes* forests; dominated by *Croton huberi* (Euphorbiaceae) and *Montanoa quadrangularis* (Asteraceae). In addition to these four dominant forest types, local plant communities are found under special circumstances, such as successional liana communities, dense stands of bamboo, and of palms. Compared to other cloud forest areas in the Coastal Range of northern Venezuela, however, the Avila cloud forests seem to be less species rich and to have less epiphytes and climbers.

Winfried Meier, *Apartado 88008, Módulo del Club Hípico, Caracas 1084-A, Venezuela. Telephone 58-2-770125, Fax 58-2-772528.*

LECTURE

Montane Forests in Protected Areas: The Ecuadorian Situation

P. Mena

Ecuador has 16 units in its national system of protected areas. A new strategy from 1989 proposes creation of 16 additional units to cover all ecosystems and offer the expected services. Many protected areas contain important extensions of montane forests and páramos in different degrees of conversion. They are: Cotacachi Cayapas Ecological Reserve, Cayambe Coca Ecological Reserve, Pululahua Geobotanic Reserve, Cotopaxi National Park, Boliche National Recreation Area, Chimborazo Faunal Production Reserve, Sangay National Park, Cajas National Recreation Area, and Podocarpus National Park. Other officially protected areas, are several watershed protecting forests with important samples of high Andean habitats (Mindo-Nambillo, Maquipucuna, Molleturo-Naranjal, Cashca, and Pasochoa). Montane forests are the most diverse but also the most altered ecosystems in Ecuador, and indeed, the protected remnants are almost the only ones that persist somehow pristine among cities, pastures, fields and eroded slopes in the Ecuadorian Sierra. These islands of forest are under severe stress from settlement, faulty official policies, and inadequate land use and distribution. Paradoxically, the least transformed forests are not protected by the legal system but by difficult access. The official agency used to be a small office in the Ministry of Agriculture, which is more devoted to production than conservation. Recently the agency was elevated to a higher level (Instituto Ecuatoriano Forestal y de Recursos Naturales Renovables, INEFAN). The production (forestry) component is still there and, as always, more powerful. The personnel are badly trained and paid people who often show a great deal of martyrdom; those in montane forests are perhaps in the most adverse condition, for living at these sites is usually quite demanding. To solve these problems a change in official government policies is needed. Large parts of the Ecuadorian conservationist community ask for creation of an autonomous, high level official institute for protected areas. The recent tendency to privatize public services should not be applied since these areas belong to all Ecuadorians and as such should be managed officially. Controlled involvement of local and international conservationist NGOs and possibly certain private profit-oriented entities (*e.g.*, ecotourism agencies) could be part of the solution. Involvement of local communities in these montane forest protected areas is fundamental, considering environmental education and the possibility of generating resources from services derived from management of protected areas.

Patricio Mena, EcoCiencia, P.O.Box 17-12-257, Quito, Ecuador. Telephone and Fax 593-2-502 409, Email: patricio@ecocia.ec.

LECTURE

The Fern Genus Elaphoglossum (Elaphoglossaceae) in the American Tropics

J. T. Mickel

Elaphoglossum is the largest and one of the taxonomically most difficult genera of ferns in the neotropics. It has over 400 species in America and 150 in the Old World. These are small to medium-sized, mostly epiphytic plants with undivided, paddle-shaped blades and acrostichoid sori. Because of the challenging number of species and their apparent similarity, they have been neglected taxonomically. The purpose of this paper is to point out the characters that are useful in the classification of the genus and describe the distribution patterns of the species in tropical America. First, the overall appearance of the plant is often diagnostic. Size of the plant, rhizome habit, stipe length, blade shape, texture, and general scaliness will frequently serve to identify the species. Most important from the standpoint of overall classification are the scales, especially those of the blade and stipe. Scales range from linear-lanceolate and entire, to toothed, to highly dissected, or further reduced to stellate hairs, trichomidia, and resinous dots. They may be pale or heavily indurated. Rhizome scales exhibit less variation, but are generally distinct from those of the frond. Characters of venation are also important, including angle of the veins to the costa, intervein distance, hydathodes, and, rarely, anastomoses. The species of Elaphoglossum are mostly of montane regions, between 1000 m and 3000 m elevation. Regions of species concentration are Mexico-Guatemala, Costa Rica-Panama, the Antilles, the Andes, the Guayana Highlands, and southeastern Brazil. Only about 30 species are widespread, here meaning that they occur in three or more of the cited regions. On the other hand, there is also a very high degree of endemism, with 3/4 of the species occurring in only one of the cited regions. The reason for the high number of local species can be traced to the fact that there is essentially no hybridization among epiphytic ferns, including Elaphoglossum, and probably no crossing between populations. Self-fertilization, resulting perhaps from gametophyte isolation on tree trunks and branches, allows mutations to be immediately fixed and microspecies to be formed.

John T. Mickel, The New York Botanical Garden, Bronx, NY 10458-5126, USA. Telephone 718-817-8636.

LECTURE

Palms of the Tropical Andes

M. Moraes, A. Henderson, H. Balslev, G. Galeano and R. Bernal

The palm flora of the tropical Andean highlands between 1000–3200 m is represented by *ca.* 90 species and 19 genera, or about 14% of the Neotropical palms. This palm flora contains mixed biogeographical elements. The lowland elements goes up along the Andean slopes not higher than 1500 m, while the typically Andean elements descend to adjacent lowland areas. Six palm genera have reached their highest diversification in the Andean region: *Aiphanes, Ceroxylon, Catoblastus, Dictyocaryum, Parajubaea,* and *Wettinia.* The origin, distribution, ecology, uses, and conservation status of Andean palms are discussed.

Mónica Moraes, Herbario Nacional de Bolivia, Casilla 10077, Correo Central, La Paz, Bolivia. Fax 5912-797511. Andrew Henderson, The New York Botanical Garden, Bronx, NY 10458-5126, USA. Henrik Balslev, Department for Systematic Botany, Aarhus University, Building 137, DK-8000 Aarhus C., Denmark. Telephone 45-86-202711 ext. 2522, Fax 45-86-139326. Gloria Galeano and Rodrigo Bernal, Instituto de Ciencias Naturales, Universidad Nacional de Colombia, Bogotá, Colombia.

LECTURE

Diversity and Phytogeography of the Asclepiadaceae from the Venezuelan Andean Forests

G. Morillo

The purpose of this paper is to give an overview of Asclepiadaceae in the Venezuelan Andean forests, their geographical and ecological distribution and their areas of concentration. Information was obtained from specimens (incl. types) and literature in Venezuelan, North American and European Herbaria, and by observations in the field. A total of 12 genera and 41 species were identified: 1 *Asclepias*, 1 *Blepharodon*, 24 *Cynanchum*, 2 *Ditassa*, 1 *Fischeria*, 2 *Gonolobus*, 1 *Macroscepis*, 1 *Marsdenia*, 5 *Matelea*, 1 *Oxypetalum*, 1 *Sarcostemma*, and 1 *Tetraphysa*. Geographic distribution of the species: Táchira 7, Mérida 12, Trujillo 5; one species widespread, from Táchira to Trujillo. Concentration of endemics in: Páramo of Tamá (5) and Batallón (2) in Táchira; Mountains of "Pueblos del sur" (6) and Sierra Nevada de Mérida (4) in Mérida; and Páramos La Cristalina and Guaramacal (2) in Trujillo. A modified Holdridge system was used for vegetation: I. Lowland and transitional forests and thickets (300–1100 m) had eight species all widespread in south America or even the Neotropics. II. Premontane and low montane forests (1200–2700 m) had 18 species, many of them with a very limited geographical distribution. III. Upper montane (dwarf) forests, subpáramos and páramos (2200–3500 m) had 15 species, most of them endemic and more than half undescribed, including a new species of *Tetraphysa*. There is a very high level of endemism in the Asclepiadaceae of the Venezuelan Andean forests: 24(58,5%) of 41 species are endemic to the area and 11 of them are limited to the premontane and low montane forests, whereas 13 are restricted to the upper parts of the mountains. *Tetraphysa* is the only typical Andean genus, representing an extension of the northern limits of the genus, previously known only from Ecuador and Colombia. Ten species are new to science. Two of them also occur in Venezuelan Guayana and two have populations in the Coastal Cordillera. Areas high in endemic species (Páramos of Tamá, Batallón, Sierra Nevada de Mérida, N and E Trujillo), coincide with the refuge areas proposed by Steyermark.

Gilberto Morillo, Fundación Instituto Botánico de Venezuela, Apartado 2156 Caracas, Venezuela. Telephone 2-662-9254, Fax 662-9081.

LECTURE

Evidence for High Biodiversity in Montane Neotropical Agaricales

G. M. Mueller

Higher fungi play an integral part in tropical forests by being intimately involved with such basic processes as nutrient cycling, nutrient uptake, and decomposition of organic matter. Little, however, is known regarding their biodiversity, host specificity, and distribution in tropical regions. Comparisons of species composition and distribution patterns for a diverse group of Agaricales (mushrooms and relatives) were undertaken to provide preliminary information on the potential biodiversity and degree of endemism in neotropical montane forests. Genera were selected from each of the three recognized suborders of Agaricales and include both saprobic and ectomycorrhizal fungi. Comparisons were made of the species composition of the selected genera from temperate North America, southern neotropical oak forests of Costa Rica and Colombia, lowland neotropical forests (especially Brazil, Venezuela, and the Lesser Antilles), and south temperate South America. The examined fungi show discrete rather than cosmopolitan distribution ranges with little overlap in species composition between these different geographic regions. Potential endemism is high in neotropical oak forests with 30–100% of the species in each of the genera studied not reported from outside the region. Finally, the ectomycorrhizal fungus/ ectotrophic host ratio for the tested genera appears to be higher in neotropical oak forests than in temperate North America. While we need data on many more genera to substantiate these findings, it is becoming apparent that the higher fungi show a high level of biodiversity in montane neotropical forests and that this diversity is not evenly distributed throughout the neotropics.

Gregory M. Mueller, Department of Botany, Field Museum of Natural History, Chicago, IL 60605-2496, USA. Telephone 312-922-9410 ext. 319, Fax 312-427-7269, Email: mueller%fmnh785.fmnh.org@uicvm.uic.edu.

LECTURE

Solanaceae in Neotropical Mountains

M. Nee

The Solanaceae have their highest generic and specific diversity in the Americas, especially in Andean South America. The diversity of the solanaceous flora of the Departamento Santa Cruz, Bolivia, is presented in relation to the major ecogeographic regions. This diversity is compared with that of central Amazon and of Estado Veracruz, Mexico. The geographic and ecological diversity of the Solanaceae parallels that of Centrospermae (subclass Caryophyllidae) in many ways, and the two groups share several physiological and morphological characteristics. It is proposed that the Solanaceae and the Centrospermae share a common ecogeographic (but not phylogenetic) origin in the southern hemisphere.

Michael Nee, The New York Botanical Garden, Bronx, NY 10458, USA. Telephone 718-817-8643.

LECTURE

Lycopodiaceae - Diversity in Neotropical Montane Forests

B. Øllgaard

Of the *ca.* 190 species of Lycopodiaceae in the Flora Neotropica area (*Huperzia ca.* 157, *Lycopodiella* 25, *Lycopodium* 8), approximately 94 (*Huperzia* 82, *Lycopodiella* 9, *Lycopodium* 3) occur mainly or exclusively in montane forest. The great majority of the species are light-demanding and sensitive to drought. The family is generally restricted to permanently humid regions or regions with a little pronounced dry season. In general terms the species of montane forests correspond ecologically to two types of habitat: 1) forest proper 2) open habitats, *e.g.* landslides, road banks in various states of vegetation regeneration. Species of the forest proper are mainly epiphytes, but a few are terrestrial in relatively light exposed forest floor habitats near the timber line. The species of open habitats are all terrestrial. The northern Andes (Venezuela to Ecuador has the greatest diversity with 54 species (*Huperzia ca.* 42, *Lycopodiella* 9, *Lycopodium* 3), followed by Central America with 45 species (*Huperzia ca.* 35, *Lycopodiella* 7, *Lycopodium* 3), the southern Andes (Peru to northern Argentina) with 37 species (*Huperzia ca.* 28, *Lycopodiella* 6, *Lycopodium* 3), Brazil (especially eastern and south-eastern) with 31 species (*Huperzia* 24, *Lycopodiella* 4, *Lycopodium* 3), West Indies with 24 (*Huperzia* 18, *Lycopodiella* 3, *Lycopodium* 3), and the Guianas with 18 species (*Huperzia* 10, *Lycopodiella* 5, *Lycopodium* 3). Brazil has the highest number of endemic species (18), followed by Central America (9), the northern Andes (7), the West Indies (3) and the southern Andes and the Guianas with two each. Compared to the species of páramos the montane forest species are generally more widely distributed with lower numbers of local endemic species.

Benjamin Øllgaard, Department of Systematic Botany, Institute of Biological Sciences, Aarhus University, 68 Nordlandsvej, DK-8240 Risskov, Denmark.

LECTURE

Patterns of Evolutionary Diversification in Brunneliaceae

C. Orozco

The Brunelliaceae is a monogeneric neotropical family, distributed widely in cloud forest of the Andean zone. *Brunellia* has close relationships with some genera of Cunoniaceae, Leguminosea, Fagaceae, and Hamamelidaceae. A phylogenetic analysis was performed in order to determine evolutionary patterns of morphological variation, as well as intrageneric relationships and the patterns of relationship with other taxa. This analysis employed the Hennig–86 program, based on floral, inflorescense, fruit and vegetative characters, and with various taxa as outgroups. This analysis identified several natural groups within Brunelliaceae. Pollen and vestiture of the leaf underside where analyzed with scanning electron microscope in 40% of the species of *Brunellia*. I evaluated the importance of the characters in the natural groups. Within each natural group variation in pollen exine was detected; the importance of this is discussed. On the other hand, variation in characters of the leaf vestiture is consistent with the natural groups.

Clara Inés Orozco, Instituto de Ciencias Naturales, Herbario Nacional Colombiano, Apartado Aereo 7495, Bogotá, Colombia. Telephone 2684336, 2693943, Fax 57-1-2682485.

LECTURE

A Remnant High-Elevation Forest in the Andes of Northern Ecuador: Floristic Composition, Structure and Diversity

W. A. Palacios, G. Tipaz and D. A. Neill

The high intermontane valleys of the tropical Andes once supported extensive forests up to timberline, but these regions have been densely populated for centuries and the natural vegetation has been almost entirely destroyed by the advance of cattle-raising, agriculture and urbanization. However, a few fragments of undisturbed forest still remain on the slopes of the interior Andean valleys, and these remnants give a hint of the original vegetation. We conducted floristic inventories in one of the few remnants of intact high-elevation forest in northern Ecuador: Loma El Corazón, in the Río Minas watershed south-east of the village of Huaca in Carchi province. Located on steep slopes of volcanic origin, the 1000-ha forest tract extends from 3100–3500 m elevation. We visited the site during different seasons of the year to make collections of the flora; we also carried out a quantitative study in the forest at 3150 m, using the method widely employed for comparative studies by A. Gentry and others: a sample of all stems ≥ 2.5 cm DBH within an area of 0.1 ha. The forest is notably tall and dense, with canopy trees up to 30 m tall and 1.5 m in diameter. Species richness is also remarkably high: a preliminary list of the flora of the site comprises 250 species including more than 60 tree species. The 0.1-ha sample included 41 species of trees, shrubs and lianas, indicating a relatively diverse woody flora for such a high elevation site, and more diverse than any known temperate-region forest. The high basal area of 7.2 m² in the 0.1-ha sample reflects the abundance of large trees. Some of the dominant tree species include *Ocotea infrafoveolata*, an undescribed species of *Ocotea*, *Weinmannia rollotti*, *W. pinnata*, and *Clusia flaviflora*. The latter species forms dense, nearly monospecific stands in some areas. Above 3600 m, this tall forest gives way abruptly to a wet páramo dominated by *Espeletia pycnophylla* and *Pentacalia*. The studies at the Loma El Corazón site suggest that tall, dense, diverse forest is the original vegetation of the inter-Andean valleys right up to timberline; the scrubby subpáramo vegetation found today in most areas at 3100–3500 m is the result of wholesale destruction of the original forest. Unfortunately, this unique remnant of undisturbed Andean forest is disappearing rapidly; the trees of El Corazón are being cut to produce charcoal and the steep slopes are being converted to croplands and pastures. The protection of this area merits top priority among conservation efforts in Ecuador.

Walter A. Palacios and Galo Tipaz, Herbario Nacional del Ecuador, Casilla 17-12-867, Río Coca 1734, Quito, Ecuador. **David A. Neill,** *Missouri Botanical Garden, P.O. Box 299, St. Louis, Missouri 63166, USA. Telephone and Fax 593-2-441592, Email: dneill@jsacha.ec .*

LECTURE

Flora and Vegetation of the Laguna de Cuicocha, in the Cotacachi-Capayas Ecological Reserve, Ecuador

M. Peñafiel

The Laguna de Cuicocha is a crater lake occupying a volcanic caldera in Cotacachi canton, Imbabura province, Ecuador, at 3100 m elevation. Within the 400-ha lake are two islands, Teodoro Wolff (41 ha) and Yerovi (27 ha). Surrounding the lake are steep to moderate rocky slopes. In the Holdridge Life Zone system, the site is classified as Montane Moist Forest. The Cuicocha lake is within the Cotacachi-Cayapas Ecological Reserve, and receives many thousands of visitors annually. The Ecuadorian national park service requested a study of the flora and vegetation of the area around Laguna de Cuicocha to aid in planning for interpretative nature trails and conservation activities. A floristic inventory was conducted in the vicinity of the lake and on the islands. Fifteen vegetation transects, each 50 × 2 m, were established around the lake and on the islands in order to characterize the different vegetation types. The transect data was analyzed using cluster analysis. A total of 410 species of vascular plants have been identified, including 22 pteridophyte, 1 gymnosperm, 93 monocot, and 293 dicot species. The analysis of the transect data revealed four principal vegetation types: 1) Grassland ("pajonal"), subject to frequent burning and other disturbance, dominated by *Calamogrostis intermedi* and *Stipa ichu*, with scattered shrubs such as *Pernettya prostrata* and *Vaccinium floribundum*; 2) Primary forest, on Teodoro Wolff island, a remnant high-elevation forest dominated by *Columellia oblonga* var. *sericea*, *Solanum* sp., and *Miconia crocea*; 3) Steep gullies with shrubby vegetation dominated by *Berberis hallii, Ageratina pseucochilca, Piper andicolum, Barnadesia,* and other shrubs; 4) Disturbed areas on the Yerovi island, strongly affected by the influx of tourists to the site, with a mixture of herbaceous and shrubby vegetation including *Stipa ichu, Arcytophyllum thymifolium, Brachyotum ledifolium,* and *Otholobium mexicanum*. The study of the flora and vegetation of the Laguna de Cuicocha will be applied to the development of educational programs for visitors to the site, and to help gain support among the general public for the conservation of the Cotacachi-Cayapas Ecological Reserve.

Marcia Peñafiel, Herbario Nacional del Ecuador, Casilla 17-12-867, Quito, Ecuador. Telephone and Fax 593-2-441-592.

LECTURE

Floristic Analysis of Andean Mountain Forests Along Altitudinal Gradients, Tachira State, Venezuela

M. Pietrangeli and V. González.

The present work was carried out in the El Valle river basin, Tachira state, Venezuela. The purpose was to characterize structure and floristic composition of the montane tropical forests. This forest type is distributed at 1900–2950 m elevation. The main variables taken to stratify the sampling procedure were: altitude, slope, exposure, land forms and kind of lithological material. This last factor because in the study area five distinct geological formations outcrop. Fifty stands were sampled in 500 m² plots. Data analysis, using multivariate techniques, separate six different kinds of forests based on the floristic composition and the relative abundance of each species present in the plot. Montane forests with greatest structural and floristic complexity were associated with soils derived from La Quinta Formation (Tropohumults), formed by a sequence of sedimentary rocks (limonite and sandstone) of Mesosoic origin. This forests type shows three tree strata. The superior tree stratum reaches 25 m and it is relatively open. Its floristic dominance is shared by *Podocarpus oleifolius*, *Alchornea triplinervia*, and *Clusia* aff. *multiflora*. The greater abundance of individuals of tree fern *(Cyathea)* is remarkable in this kind of forest. The basal area can reach 70 m²/ha and the tree species richness varied from 35– 40 species/ha. Elfin Forest, associated to mountain slopes above 2800 m elevation, was found in the other extreme of the altitudinal gradient studied. These communities had one tree stratum which varied from 4–8 m in height. Tree stems are very gnarled and branched, their leaves are coriaceaus, microphyllous in size, and evergreen. The dominant species are: *Hedyosmum glabratum*, *Weinman-nia fagaroides*, *Symplocos rigidissima*, *Geissanthus andinus* and *Palicourea venezuelensis*. Most of the stems are totally cove-red by a moss layer of different species. The basimetric area varied from 15–20 m²/ha and the tree species richness is not higher than 15 per hectare. Between these two contrasting kinds of forests, it is possible to recognize four additional ones, based on the physiognomy and floristic composition. The floristic analysis of the 50 stands studied allows us to recognize 153 species grouped in 84 genera and 41 families. The genera most rich in species are: *Myrcianthes*, *Miconia* and *Geissanthus*, while at the family level the richest are: Melastomataceae, Lauraceae, and Myrcinaceae. Project supported by CONDES (870-91), Univ. del Zulia; Centro Ecologico de las Tierras Altas, El Cobre, Táchira and FUNDACITE-Táchira.

Miguel Pietrangeli, Univ. del Zulia, Fac. de Ciencias, Dpto. de Biología, Apdo. 526, Maracaibo 4011, Zulia, Venezuela. Valois González, Univ. Central de Venezuela, Fac. de Ciencias, Inst. Zoología Tropical, Apdo. 47058 Caracas, Venezuela.

LECTURE

Montane Forests of the West Indies

G. R. Proctor.

The West Indies are customarily classified in three main island groups: the Bahamas, the Greater Antilles, and the Lesser Antilles. Only the latter two groups support montane forests. Of all the neotropical montane forests, those of the West Indies are probably the best known, as shown by the scope and thoroughness of modern publications, and the intensity of field activity in what are relatively small areas. However, much habitat destruction has occurred, especially during the past two centuries, and is now continuing at proportionately accelerated rates on most islands with the possible exception of Puerto Rico. In view of high rates of local endemism, there is a real danger of multiple species extinctions in the near future. Comparative diversity of selected localities is discussed with particular reference to ferns and a few other taxa, but it is emphasized that the just-beginning "Flora of the Greater Antilles" project will be necessary for a balanced phytogeographic knowledge of the region.

George R. Proctor, Department of Natural Ressources, Puerto Rico.

LECTURE

Diversity of the Flora, Vegetation, and Environmental Gradients in the Andes of Colombia

J. Orlando Rangel-Ch.

The purpose of this study was to characterize the vascular plant diversity level (as species per family and genus) in different life zones along a mountain gradient in Colombia. The study area was Sierra Nevada de Santa Marta; Parque Natural de los Nevados; Macizo Central (Valle del Magdalena–Volcán del Puracé), and Macizo de Tatamá (Cordillera Occidental). In two zones, characteristic of the mountain gradient, ecological series were determined based on distribution of vegetation types. The two zones were the upper region or páramo and the base or tropical region. In the páramo region the most abundant families, in terms of species numbers, were: Asteraceae, Gramineae (Poaceae), Polypodiaceae, Ericaceae, and Rosaceae. In the Andean region the order of importance was: Asteraceae, Orchidaceae, Polypodiaceae, Rubiaceae, and Ericaceae. In the sub-Andean region the dominant families were: Rubiaceae, Leguminosae, Melastomataceae, Orchidaceae, and Asteraceae. In the tropical region the dominant families were: Leguminosae, Rubiaceae, Piperaceae, Asteraceae, and Sapindaceae. Other interesting phytogeographic features are: 1) the families Rubiaceae, Asteraceae, Melastomataceae, Solanaceae, and Piperaceae are distributed in various altitudinal zones and the three first of them even reach species dominance in the páramo zone, 2) along the mountain gradient various genera tend to have high species numbers; *Miconia, Piper, Peperomia, Anthurium, Inga, Psychotria,* and *Palicourea* are genera with broad thermic adaptability, 3) according to the predominant growth form in each of the dominant families there is an equilibrium between families dominated by herbs (Poaceae, Asteraceae, Polypodiaceae) and families dominated by woody growth forms (Rubiaceae, Melastomataceae, Leguminosae). On the top of the mountain gradient variations in topography and ecoclimate are correlated with community physiognomy and zonal distribution of the vegetation; the lower parts are dominated by thickets, brushwood, and stunted forests, the higher parts are dominated by meadows. Variation in precipitation produces segregation of páramo communities varying from very humid ones with over 3000 mm precipitation annually to very dry ones with about 700 mm annual precipitation. In the lower parts variations in annual quantities and distribution of precipitation creates an ecological series ranging from semi-deserts with less than 1000 mm rain to pluvial forests with over 3000 mm annual precipitation.

J. Orlando Rangel-Ch., Instituto de Ciencias Naturales, U. N. de Colombia, Apartado 7495, Santafe de Bogotá, Colombia.

LECTURE

Conservation of Biological Diversity in a World of Use

K. H. Redford

Biological diversity conservation is one of the watchwords of this decade. There has been much discussion about how to conserve biodiversity and a great deal of money spent pursuing this goal. Despite this, in almost all programs designed towards such goals, the meaning of "biological diversity" is rarely made explicit and the word "conservation" is either left undefined or defined as "sustainable use" as is done in the influential document, "Caring for the Earth." In this talk I suggest that using most definitions of biological diversity combined with this meaning of conservation, the term "conservation of biological diversity" is an oxymoron. We must return to distinguishing between "conservation" and "preservation" in order to achieve the goal of maintaining biodiversity.

K. H. Redford, Program for Studies in Tropical Conservation, Center for Latin American Studies, University of Florida, Gainesville, Florida 32611, USA. New address: Director of Conservation Science and Stewardship, Latin America Division, The Nature Conservancy, 1815 N. Lynn Street, Arlington, Virginia, USA

LECTURE

Asteraceae in the Neotropical Montane Area with an Emphasis on the Andes

H. Robinson, V. Funk, G. McKee and J. Pruski

The Asteraceae is the largest family of flowering plants with over 20,000 species. In the Andes above 1000 m, there are *ca.* 2900 species in 260 genera. The Asteraceae in the Andes range from ephemerals to 18 meter tall trees and are found in all habitats from forests to páramo and puna where some species grow at the foot of glaciers. The family is one of the dominant elements above tree line. Species can be apomictic, self-pollinating, out crossing or dioeceous and pollination is by wind, a variety of insects, or by hummingbirds. Distribution of species in the family is best analyzed by individual tribes, some of which are larger than many plant families. In the Andes the tribal distributions fall into one of four groups. The most important tribes in the area are the Heliantheae, Eupatorieae, Mutisieae, Liabeae, and Barnadesieae, which are not only primarily Western Hemisphere in distribution but also probably originated there. The tribes Senecioneae, Astereae, and Vernonieae are well represented in the Andes but are common in both hemispheres. The Plucheae, Gnaphalieae, and Lactuceae are represented by a few endemics and widespread species but their distribution is primarily nearctic and eastern hemisphere. The remaining tribes are less relevant to the origin and evolution of Andean Asteraceae. Detailed analysis of the species composition in the Andes show contrasting patterns for different tribes. Several overall patterns have emerged. For instance, some taxa, such as the Liabeae, originated in the Andes and spread to other areas. Other taxa such as *Viguiera* and *Verbesina* of the Heliantheae, originated in central Mexico or Central America but have radiated extensively in the Andes. Also, a number of groups show a definite tendency to have most of the species diversity concentrated either from Ecuador north or from Peru south. The geologic history of the Andes has naturally played an important role in the evolution of the Asteraceae and a definite pattern can be seen based on the fact that the northern Andes have more recently obtained high elevation. This is evident in the opportunistic Asteraceae that have moved into this area such as *Espeletia*, from lower elevations and *Gynoxys* from further south. Because they are such an important part of the Andean flora particular tribes of the Asteraceae such as the Eupatorieae, Liabeae, Heliantheae, Mutiseae, Barnadesieae, and Senecioneae furnish excellent examples of the probable history of the evolution of this flora over the last 20 million years.

Harold R. Robinson, V. A. Funk, Gregory S. McKee and John F. Pruski, U.S. National Herbarium, Department of Botany, National Museum of Natural History, Smithsonian Institution, Washington D.C. 20560, USA. Telephone 202-357-2534, Fax 202-786-2563, Email(BITNET): MNHbo003@SIVM (for VAF).

LECTURE

The Montane Cloud Forests of Venezuelan Andes

H. A. Rodríguez-Carrasquero

The Venezuelan Andes are located in the Western part of Venezuela, in northern South America. It starts at the Colombian border (Tamá Páramo) through the States of Táchira, Mérida and Trujillo, up to Lara depression, and it extends to part of Barinas, Portugues, and Lara States. It has a length of 450 km, it covers a total area of 36,120 km^2, and ranges from 7°30'–10°10'N, 69°20'–72°50'W. The montane forests are distributed through the main watersheds of the region (Uribante, Chama, Santo Domingo, and Motatan rivers) in the central part of the Andean region, the northern area of Maracaibo Lake watershed, and the southern part of the Andean foothills, limits with the Venezuelan lowlands. Some parts of Venezuelan Andes are protected as national parks or hydraulic reservoirs. Plots of one ha (250 × 40 m) were marked; field data consisted of common names, trees with more than 20 cm dbh, commercial and total height. The floristic composition of the Venezuelan Andes showed that the most important families were Euphorbiaceae, Lauraceae, Burseraceae, Lecythidaceae and Sapotaceae, but in few other plots Podocarpaceae, Cunoniaceae, Myrtaceae and Lauraceae were dominant.

Henry A. Rodríguez-Carrasquero, Forest Management Department, Forestry School and Silviculture Institute, The Andes University, Mérida, Venezuela.

LECTURE

Rosaceae in the Ecuadorean Montane Forest

K. Romoleroux

Currently 13 genera and about 60 species of Rosaceae have been recorded in Ecuador. Although most genera are found in the upper montane forest and páramos, the family occurs also in the lower montane forest. The genera of Rosaceae which are most common in the montane forest are *Rubus, Prunus,* and *Hesperomeles,* others as *Acaena* and *Polylepis* have some species growing in this area, and also naturalized genera as *Crataegus, Fragaria,* and *Duchesnea* occur in this region. The genus *Rubus* with at least 17 species in Ecuador, mainly occurs at 2000–3000 m elevation, but some species have been found growing at elevations of 500–800 m. *Prunus* is represented in Ecuador by about five species, with a wide altitudinal range from 300–3000 m. *Hesperomeles* with two species and three varieties is a common genus of the Ecuadorean montane forests, especially at elevations of 2500–3500 m. *Acaena* and *Lachemilla* mainly occur in the páramos but some species descend to 2500 m. *Polylepis* occurs in the highest part of the Andean Ecuadorean region as patches of forests, nevertheless some species as *P. reticulata* and *P. lanuginosa* have been recorded from 2800 m. The genera *Crategus, Fragaria,* and *Duchesnea* are represented by one species each in Ecuadorean montane forest at elevations of 2000–2500 m. The montane forest is a very important ecosystem for the variability and development of the Rosaceae in Ecuador.

Katya Romoleroux, Pontificia Universidad Católica del Ecuador, Departamento de Biología, Herbario QCA, Apartado 17-01-2184, Quito, Ecuador. Telephone 529-250 ext. 279, Fax 593-2-567117.

LECTURE

Fungi of Antioquia, Colombia

Y. Saldarriaga Osorio

The department of Antioquia covers 63,612 km^2; because of its broken topography, it includes a number of different climates and altitudinal zones, and therefore harbours a diverse mycoflora. There are only slight changes of temperature during the year but there are two rainy seasons. The pronounced differences in elevation, however, produce temperature zones along the altitudinal gradient Due to the large extension of the department, a number of representative areas were selected as collecting sites for this study. These areas represent a number of life zones varying from very humid to dry tropical forests and from lowland over premontane to low montane forests. In total 1405 collections of macromycetes were made, representing 34 families and 102 genera. The most diverse families were the Tricholomataceae with 22 genera, Polyporaceae with seven, and Agaricaceae with six genera, respectively. The genera most frequently collected were: *Marasmius* (13.6%), *Lepiota* (4.7%), *Xylaria* (3.3%), *Coprinus*, and *Pleurotus* (2.8%). The following species were reported for the first time in Colombia: *Tricholoma cystidiosum*, *Cordyceps capitata*, *Cordyceps ophioglossoides*, *Phaecollybia atenuata* subsp. *mexicana*, *Ripartella brasilliensis*, *Geastrun finbriatun*, *Licoperdon curtissii*, *Aguascypha hydrophora*, *Favolous tesselatus*, *Lentinus velutinus*, and others. The decaying wood fungi are: *Aguascypha hydrophora*, *Auricularia polytricha*, *Auricularia judea*, *Cookeina tricholoma*, *Cookeina brassiliensis*, *Lentinus crinitus*, *Lentinus strigosus*, *Pleurotus ostreatus*, *Polyporus sanguineus*, *Polyporus trichomallus*, *Xylaria hypoxilon*, and *Caripia foetidum*. There are 26 edible fungi, but none of them are used, due to the little knowledge in our environment. The toxic fungi found were: *Amanita muscaria*, *Coprinus plicatilis*, *Coprinus disseminatus*, *Coprinus atramentarius*, *Coprinus niveus*, *Russula emetica*, and *Stropharia semiglobata*. Within the group of parasitic species we have: *Cordyceps capitata* and *Cordyceps ophioglossoides* found parasitizing fungi such as *Elaphomyces granulatus* and *Elaphomyces reticulatus*, respectively. Parasites of trees were the following: *Amauroderma omphalodes*, *Auricularia polytricha*, *Polyporus hydnoides*, *Ganoderma aplanatum*, *Polyporus arcularia*, etc.

Yamille Saldarriaga Osorio, *Universidad de Antioquia, Departamento de Biologia, Apartado 1226, Medellin, Colombia. Fax 263-8282.*

LECTURE

Restoration: The challenge for Conservation in Tropandean Landscapes of Ecuador

F. Sarmiento

For tropical mountains in Ecuador, where a shift in policy needs to refocus the "setting-aside" of protected areas to include a more accurate small-area strategy of adaptive management, including natural areas bordering agricultural fields, and improvement of reserves that are already present. I call attention to the inter-Andean forest contrasting it with the trans-Andean forest. The cloud forest belt of the western and eastern slopes of the Andes of Ecuador, retains substantial forest cover because climate and topography slowed down deforestation on the steep slopes. Penetration for grazing and timber production started in the 1970s; secondary roads were built for colonisation and agricultural production, initiating the decline of what was thought to be pristine primary montane forest, without noticing the presence of indigenous tribes (*i.e.*, Quijos, Yumbos) who in big numbers populated the mountain region. The tropandean bioma, however, also included the inter-Andean valleys located in small watersheds among the big parallel ranges, offering a mosaic of habitats that can be considered unique to the equatorial Andes region. Some of them shows warm aridity, often driven by phoen effects on the rain-shadow areas, while some of them are quite humid and cold seepage basins. This offers a wide range of microclimates that dictate the establishment and development of a diverse biota. Being a mesic environment, the area supported the majority of anthropogenic modification since pre-Inca times. Indeed, the oldest human settlement in Ecuador is located among Los Chillos and Tumbaco valleys. The area has been over-exploited and its natural plant cover is almost lost. Small remnants survive in remote gorges or isolated ridges and in the brooks and streams near urban areas. The rural landscape has regrettably been dominated with exotics. No critical area has been left without human intervention. Therefore, the approach of restoration ecology is urgently needed in order to establish realistic conservation goals for tropical mountains. It is crucial to ensure survival of inter-Andean forests and sustainable agriculture in the Andes. With a landscape ecological approach, a proposal for the upper Guayllabamba River basin is presented emphasizing the rural appraisal technique and determination of habitat suitability for restoration ecology practices. A call is being made to the planners and decision makers to consider new strategies requested from gap analysis of the basis, such as a riverine corridor for the tributaries that connects core reserves as well as buffering small reserves to integrate a bigger area in the management of rural development.

Fausto O. Sarmiento, The Institute of Ecology, University of Georgia, 102 Ecology Building, Athens, Georgia 30602-4022, USA. Telephone 706-542-2968, Fax 706-542-6040.

LECTURE

Maquipucuna: A Way to Conserve Tropandean Landscapes

F. Sarmiento

The Ecuadorian Foundation for Conservation MAQUIPUCUNA, (Quechua for "the caring hand") is a non profit organisation registered under Ecuadorian law to promote study and conservation of nature. Founded in 1987, it took the effort of many professionals in different disciplines to prepare an organisational agenda which include research of biotic resources, management of privately-owned reserves and environmental education at large. With the financial assistance and the technical expertise of different organisations, both inside the country and overseas, Maquipucuna has successfully achieved its primary goal of creating a framework for conservation with the bottom-up approach. Maquipucuna has already started a series of programs dealing with community forestry and management of sensitive areas. In particular, the LA PAZ project, in the buffer area of the Podocarpus National Park, has proved successful in integrating the community with the maintenance and enhancement of the remaining forests. Fostering of the research agenda is crucial. The newly created "Thomas Davis" Research Station for Cloud Forest Ecology is already generating basic information of biological inventories of one rich area in the mountains of Maquipucuna near Nanegalito and Nanegal, in NW Ecuador, one of the "hot spots" (sensu Myers) for biodiversity in the world. The reserve itself encompasses 3000 ha of native forest with some abandoned fields and old agricultural farms in the surrounding, which is going to be managed as a buffer zone for the protected area; also, research projects of forest regeneration in the montane forest are being prepared in the nearby area including development of tropical forestry assays, nurseries for native species, translocation, soil fertility and nutrient cycling and long term monitoring of cloud forest processes, including progressive succession and retrogression as well as the use of spatial data for landscape studies. An invitation to all Neotropical montane forests ecologists is presented to contact the Foundation headquarters in Quito as well as the field sites that are being implemented in particular the "Thomas Davis Research Station" in the Maquipucuna reserve.

Fausto O. Sarmiento, The Institute of Ecology, University of Georgia, 102 Ecology Building, Athens, Georgia 30602-4022, USA. Telephone 706-542-2968, Fax 706-542-6040.

LECTURE

Ethnobotany of the Chachapoyas People. Use of Plants from the Peruvian Montane Forest and Related Areas: An Anthropologist's View

I. Schjellerup and A. Sørensen

This paper gives information on the geographical setting and vegetation of today in the southern part of the Province of Chachapoyas, Department of Amazonas in Peru. Ethnographic data and historical archival studies allow us to form a picture of the use of plant material in the culture and the recent changes. Of special interest is the utilization of the different ecological zones in the mountains and the use of medicinal plants. The picture is of a vulnerable landscape on which more emphasis should be put in the future.

Inge R. Schjellerup, Depart. of Ethnography, National Museum of Denmark, Ny Vestergade 10, DK-1471 Copenhagen K, Denmark. Telephone 45-33-134411 ext. 368, Fax 45-33-144421. Anne Marie Sørensen, Inst. of Botany, The Royal Vet. and Agricult. University, Copenhagen, Denmark. Mailing address: Bygårdstræde 7, 1.tv, DK-2700 Brønshøj, Denmark. Telephone 45-31-600245.

LECTURE

Floristic Exploration and Phytogeography of the Cerro del Torrá (Chocó, Colombia)

P. Silverstone-Sopkin and J. Ramos-Pérez

The purpose of three expeditions to the Cerro del Torrá was to expand knowledge of the composition of its plant communities and to ascertain whether its flora represents a phytogeographical entity distinct from that of the Cordillera Occidental of the Andes. The Cerro del Torrá (2770–2800 m), in the Departamento del Chocó, Colombia, is an isolated mountain west of the Cordillera Occidental. Data were obtained from CUVC, MO, NY, and US. Four vegetational zones can be recognized between 1600 m and the summit; the highest of these is a páramo without *Espeletia*. Collections of vascular plants include 468 species (84 pteridophytes and 384 spermatophytes). Few families dominate the angiosperm flora. The most species-rich family is Orchidaceae. The six most diverse families include half the collection numbers and species of angiosperms, and the 12 most diverse families include over two-thirds of the numbers and species (even though 74 angiosperm families are present). Ferns, because of their efficient wind dispersal, are much more widely distributed than angiosperms and thus show less endemism. Distributional data support Gentry's assertion that the floristic difference between lowland and montane forests seems less clearly demarcated on the Pacific than on the Amazonian side of the Andes; 36.8% of identified Torrá vascular plants occur in Chocoan lowlands, but only 19.3% occur in Amazonian lowlands. Of the vascular plants, 14.4% are widespread in the Neotropics, 14.7% are widespread in the Andes, 46.9% are widespread in the northern Andes only (Venezuela–Ecuador), and 25.2% are endemic to Colombia. Only 24.2% of the vascular plants occur also on the nearby Cerro del Ingles in the Cordillera Occidental. There is an overwhelming phytogeographical affinity with the Colombian Andes (85.3% of the vascular plant species are found on at least one of the three Colombian cordilleras) and in particular with the Cordillera Occidental of Colombia (74.8%). There is a low percentage of narrow endemism (only 5.2% of the species are restricted to the Torrá). The data suggest that the flora of the Cerro del Torra is not a phytogeographical entity distinct from the flora of the Cordillera Occidental. Nevertheless, the flora of the Cerro del Torrá is of unusual interest. The low percentage of narrow endemism may be an artifact of the paucity of identifications, particularly in orchids. The monotypic family Alzateaceae has been collected in Colombia only on the Cerro del Torrá. Of identified vascular plants, 20.9% are rare and 56.4% are not included on the Chocó checklist of Forero and Gentry, and most of these are new records for the Department of the Chocó.

Philip A. Silverstone-Sopkin, Depto. de Biología, Universidad del Valle, A. A. 25360, Cali, Colombia. Telephone 393243, Fax 57-923-392440 or 57-923-396120.

LECTURE

The Lichen Flora of the Colombian Montane Forests

H. Sipman

To give an impression of the lichen flora of the neotropical montane forests, results from one of the better-known parts, the Colombian Andes, are presented. The montane zone is defined as ranging from 800–3200 m, and all epiphytic species are considered as forest taxa, except when indications of the contrary are present. The sources of information were literature records and unpublished data from herbarium specimens in Berlin (B). Records which have not been confirmed by recent revisions are considered questionable, because the taxonomy of lichens has undergone considerable changes in recent years by the application of new characters and improved methods (*e.g.*, chemistry, ascus structure). About 700 lichen species, in 140 genera, are presently recorded. For only about 40% of the genera recent revisions are available, and over 50% of the recorded species are questionable, including most species of such conspicuous groups as *Lobaria, Leptogium, Sticta,* and *Usnea.* The result indicates that the montane zone is richer in lichens than the lowlands or the páramo zone. Many species are represented only by nineteenth century collections (A. Lindig!) and have not been found since. This seems to reflect the poor taxonomical knowledge of most lichen groups of the montane forests: only last century lichenologists felt confident enough to name them. Moreover the large-scale decrease of montane forests has made collecting more difficult and recent collections from primary montane forest are scarce as compared with the last-century ones. Evaluation of the unquestioned records suggests that the majority of the lichens are widely distributed in the Neotropics. Pantropical taxa may be almost as common, and only few taxa appear to be confined to the Andes, especially the northern part of it.

H. J. M. Sipman, Botanischer Garten & Botanisches Museum, Königin-Luise-Str. 6-8, D-1000 Berlin 33, Germany. Telephone 030-83006149, Fax 030-83006218.

POSTER

Ethnobotany of the Chachapoyas People. Use of Plants from the Peruvian Montane forests and Related Areas: A Botanist's View

A. Sørensen and I. Schjellerup

Ethnobotanic research concerning plants from the Chachapoyas area was carried out in 1987, and was a part of a more extensive anthropological project. The study area concentrates on the peasant community of Chiquibamba situated in the NE Peru. This isolated mountain region is of great interest because the area is poorly studied and contains a unique vegetation belonging to the ecological transition zone between the more humid Andes in the North (páramo), and the drier parts in the south (puna). Plant specimens were collected in the montane forests of the valleys (2700–3400 m) and in the jalca zone with its remaining stands of high altitude forest (up to ca. 4000 m). The ethnobotanical research was concentrated on medicinal plants, but also the use of dying plants, edible fruits, and wood for construction and implements was registered with a total of more than 300 plants. Historical sources give information about the Chachapoyas People as important "curanderos" (healers) and still today the curandero is important for the local communities. A continuos trade of medicinal plants between different ecological climatic zones takes place. The montane forest and its related areas deserve as much attention as the tropical rain forest because they too are in a great danger being destroyed because of the demand of land for agriculture and cattle, and the demand for firewood. When the forest disappears, nature resources, which are so important to an isolated and (nearly) self-sufficient community, disappear too. And with it, an unique flora.

Anne Marie Sørensen, Inst. of Botany, The Royal Vet. and Agricult. University, Copenhagen, Denmark. Mailing address: Bygårdstræde 7, 1.tv, DK-2700 Brønshøj, Denmark. Telephone 45-31-600245. Inge R. Schjellerup, Depart. of Ethnography, National Museum of Denmark, Ny Vestergade 10, DK-1471 Copenhagen K, Denmark. Telephone 45-33134411 ext. 368, Fax 45-33144421.

LECTURE

Using Reserves as In Situ Conservation Sites for Wild Crop Relatives

C. Sperling, R. Castillo, S. Keel, M. Gavilanez, P. Yánez and C. Hernández

Besides preserving biodiversity, protected areas in the Andes have an added benefit to agricultural scientists by conserving wild crop genetic resources *in situ*. However, there has been little data on the actual extent to which wild relatives of crop plants occur in protected areas or the significance of those species known to occur. To support the hypothesis a survey of existing herbarium records and a field inventory for congeneric species of crop relatives was conducted in three protected areas of Ecuador: Cayambe-Coca Ecological Reserve, Podocarpus National Park, and Machalilla National Park. Results thus far indicate that numerous wild relatives in the secondary and tertiary genepool are present in the reserves. Wild relatives of crop species in *Arpium, Arracacia, Carica, Cucurbita, Fragaria, Lupinus, Passiflora, Physalis, Ribes, Rubus, Solanum, Vaccinium* and other genera are documented. Most of the wild species identified are relatives of temperate horticultural crops. Germplasm samples of representative species have been placed in *ex situ* storage facilities of INIAP where they are available to plant breeders for experimentation. It is expected that the genetic diversity of only a small portion of the wild crop relatives in the reserves can be adequately preserved by other than *in situ* conservation. The authors also suggest that the assessment of wild crop genetic resources be considered when new conservation sites are being proposed.

Raul Castillo, INIAP, Quito, Ecuador. Current Address: University of Wisconsin, Dept. of Horticulture, 1575 Linden Dr., Madison, Wisconsin 53706, USA. Telephone 608-262-7406, Fax 608-262-4743. Mauricio Gavilanez, Patricio Yánez, Consuelo Hernández, Pontificia Universidad Católica del Ecuador (PUCE), Herbario, Apartado 2184, Quito, Ecuador. Telephone 593-2-529240 ext. 279. Fax 593-2-567117. Shirley Keel, Latin American Science Program, The Nature Conservancy, 1815 North Lynn Street, Arlington, Virginia 22209, USA. Telephone 703-841-2714, Fax 703-841-1283. Calvin R. Sperling, U.S. Dept. of Agriculture, National Germplasm Resources Laboratory, Bldg. 003, BARC-West, Beltsville, Maryland 20705, USA. Telephone 301-504-5251, Fax 301-504-6305.

POSTER

Aspects of Vegetation Diversity amidst Cloud Forest Systems on Guaramacal Mountain, Trujillo State, Venezuela

B. Stergios and N. Cuello

Descriptive studies and on-site evaluations of vegetation diversity were carried out in a low-altitude (2400 m elevation) cloud forest of Guaramacal Mountain, Cruz Carrillo National Forest in Trujillo State, Venezuela, approximately 9°16'N; 70°12'W. Special attention was given to floristic composition of understory and canopy trees, vertical and horizontal stratification structure and general observations concerning vegetation diversity. Field study methods employed for evaluation consisted of the one-tenth hectare, 50 × 2 m transect procedure currently in use for forest system diversity studies, where tree individuals with dbh of 2.5 cm or greater were considered. Data generated by the ecological parameters tree density, frequency, abundance, and importance value index for each species encountered in the sampled area are presented, as well as the number of families, genera and species represented. Tree structure profiles were constructed employing two artificial stratification groupings: canopy trees from 8–25 meters tall and lower, understory trees from 3–8 meters in height. Thicket vegetation consisting of herbs and shrubs was found to be quite dense, consisting primarily of such shade-loving taxa as: *Psychotria* and *Palicourea* (Rubiaceae), *Geonoma* and *Chamaerodea* (Palmae), *Chusquea* (Gramineae), and the fern genera *Asplenium, Blechnum, Diplasium, Elaphoglossum* and *Thelypteris*. Young saplings of canopy tree species were also observed to be quite abundant. As a result of the plant diversity analysis undertaken, *Beilschmiedia, Nectandra, Ocotea* and *Persea* (Lauraceae); *Miconia* and *Ossaea* (Melastomataceae); *Eugenia* and *Rudgea* (Rubiaceae); and *Cybianthus* and *Myrsine* (Myrsinaceae) were found to be most dominant. A species inventory of cloud forest floristic composition is presented, as well as their relative position within the cloud forest belt of the Cruz Carrillo National Forest. The information obtained provides a valuable tool for future study and conservation efforts in the upper montane Venezuelan Andes.

Basil Stergios and Nidia Cuello, University Herbarium, Programme of Renewable Natural Resources, UNELLEZ, Mesa de Cavacas, Portuguesa State, Venezuela 3323.

LECTURE

Vegetation Reorganisation on Earthquake-triggered Landslide Sites in the Ecuadorian Andes

M. J. Stern

The long history of frequent landslides has contributed to the composition and structure of the Andean cloud forests, especially on steep slopes. I studied reorganisation of vegetation on landslide scars resulting from the earthquakes and unusually high rainfall of March, 1987, in NE Ecuador. The study site was located in lower montane forest (1500 m) on the Quijos River watershed, Napo Province. The point-quarter method on linear transects and small stratified quadrates were used to sample the vegetation in disturbed (landslide) sites and in the adjacent undisturbed forest. The landslides were divided into zones (steep upper slope, lower debris fan, flood plain) with different physical factors that might contribute to patterns of plant community development. Rate of seedling establishment and sprouting on the landslides was variable in space and time. The first herbaceous plants (*e.g., Blechnum, Equisetum*) and shrubs (*e.g., Piper, Baccharis, Senecio, Miconia*) appeared in the mounds of soil and debris at the foot of the landslides. A climbing bamboo (*Chusquea* sp.) appeared on a landslide site two years post-earthquake, probably resprouting from an intact rhizome. On the steepest slopes, protected gullies and isolated patches of vegetation that survived the massive soil slumps and uprooted trees provided favourable microhabitats for early establishment by colonisers. *Tessaria integrifolia* dominated the flood plain one year after the earthquakes. The number of individuals of this fast-growing tree species per unit area was negatively correlated with distance from the river ($p <0.001$), indicating its importance in the riparian habitat. Other flood plain colonisers, such as *Gynerium sagittatum* and various species of *Piper*, became more important four years later. Cluster analysis showed that the species composition of the steep upper portion of the landslides was most similar (0.76) to the undisturbed forest and most dissimilar (0.18) to the flood plain. The relatively narrow upper slide's location near to the undisturbed forest suggests that proximity to a pool of potential colonisers is important to the initial species composition of the vegetation on these partially denuded landslide sites. The series of earthquakes in 1987 was not a unique event in the abrupt topography of the Ecuadorian Andes. The frequent movement of soil on these unstable slopes has influenced the selection for aggressive pioneer plant species and contributed to the heterogeneous structure of the montane cloud forest. The earthquake-triggered landslide disturbances are among the factors that have created and maintained the high floristic diversity in the upper Amazonian forests.

Margaret J. Stern, Arnold Arboretum, Harvard University, 22 Divinity Ave., Cambridge, MA 02138, USA.

LECTURE

Phytogeography of Premontane and Montane Gymnosperms

D. Wm. Stevenson

There are only a few species of gymnosperms found in montane regions of the Andes. These belong to two very distinct, unrelated genera, *Podocarpus* (Pinales, Podocarpaceae) and *Ephedra* (Gnetales, Ephedraceae). The former is cosmopolitan in the southern hemisphere and is found from lowland tropical to cool wet temperate habitats. *Ephedra* is a cosmopolitan genus (except for Australia) and is found in habitats with a pronounced dry season that generally corresponds to that found in a Mediterranean climate. The general climatic and edaphic conditions found in Andean premontane and montane habitats would seem appropriate for most temperate conifer genera. Because there are no lowland neotropical gymnosperm genera except cycads, it is assumed the montane Andean gymnosperms are not the result of *in situ* development, but rather represent dispersal events. While most conifers and species of *Ephedra* are wind dispersed, the Andean montane species appear to be animal dispersed because they have brightly coloured fleshy seeds. In *Ephedra*, the red fleshy layer that encloses the seed is derived from a pair of subtending bracteoles whereas in *Podocarpus* the purple to silver fleshy layer is a true sarcotesta. Andean species of both genera have a thick, hard, resistant sclerotesta. Several, but not all species in these genera, have brightly coloured seeds, and it is likely that the Andean premontane and montane species of *Ephedra* and *Podocarpus* are related to them.

Dennis Wm. Stevenson, New York Botanical Garden, Bronx, NY 10458, USA.

LECTURE

Andean Symplocaceae

B. Ståhl

The family Symplocaceae is generally considered to include a single genus, *Symplocos*, and about 250 species divided more or less equally between Asia and the Americas. The majority of the New World Symplocaceae are Andean, being prevalent in moist or wet forests between 2000 and 3500 m altitude. A comprehensive and modern treatment of the family in South America is lacking. However, a revision of Andean Symplocaceae is well under way and is being published country by country. At present the Symplocaceae in western and north-western South America include 75 species, 35 of which are still unrevised. Three or four species belong to subgenus *Hopea*, the remaining to subgenus *Symplocos*. Many new species have been described, especially from Ecuador, indicating that the species number probably will increase as more material becomes available. Although the work is far from being completed, some features with regard to diversity and distribution of Andean Symplocaceae begin to emerge. Firstly, the number of narrow endemics appears to be high. Secondly, species occurring over larger areas usually seem to have patchy distributions (*e.g.*, southern Venezuela–central Colombia–southern Ecuador–northern Peru). Thirdly, some areas of the Andes (*e.g.*, southern Ecuador) are very rich in Symplocaceae, whereas others (*e.g.*, northern Ecuador) have but a few species, if any. The patterns described above are partly real and partly collecting artefacts. Shrubs and trees of *Symplocos* are usually quite anonymous, especially in fruit, and are no doubt often neglected by collectors, and many of their natural habitats may have vanished long before they even had a chance to become collected. In fact, Andean Symplocaceae are poorly represented in the herbaria and many narrow endemics are likely to have wider distributions than is shown by extant records. At the same time, however, intensive collecting in different areas with montane, Andean forests have yielded widely different results as to the number of species of *Symplocos* found (figures vary from nil to eleven).

Bertil Ståhl, Botanical Institute, University of Stockholm, S-106 91 Stockholm, Sweden. Telephone 46-8-161215, Fax 46-8-165525.

LECTURE

Cretaceous and Tertiary Paleobiogeographic History of Andean Angiosperm Families

D. W. Taylor

To understand the current biogeography of Andean angiosperms, one needs a historical perspective. The present distribution of angiosperms is based on current ecological and climatic factors that are laid over these historical aspects of mountain building, geographic connections to Antarctica-Australia, Africa and Meso-North America, and paleoclimate. I review the geologic history of the Andes including mountain building, continental positions, and paleoclimate. During the past 130 million years, mountain building appears to have been episodic with progressive movement to the east. Interchange has been possible between South America and Africa until the mid-Cretaceous and to Antarctica-Australia and Meso-North America until the end of the Cretaceous and again during the Neogene. Finally, similar climatic zones appear to have existed throughout this time, although there were shifts in their specific positions. Within this framework I examine the fossil record of north Andean and south Andean based families. These records are compared to the records from Antarctica-Australia, Amazonia, and Meso-North America. These clearly show the South Andean elements existed there by the late Cretaceous while the North Andean elements were there by the mid-Tertiary. Most of both Andean paleoflora are composed of Andean or Amazonian-centered families and although the south Andean flora shares taxa with the Antarctica-Australian paleoflora as does the north Andean flora with paleoflora from Meso-North America. This suggests that the Andean floras have had a long history and the geologic changes over time may have caused high levels of speciation, endemism, and complex modern biogeography.

David Winship Taylor, Department of Biology, Indiana University Southeast, 4201 Grant Line Road, New Albany, IN 47150, USA. Telephone 812-941-2377, Fax 812-941-2637.

LECTURE

Evolution and Biogeography of Pleurozia (Hepaticae; Pleuroziaceae)

B. M. Thiers

Pleurozia is a small genus of leafy hepatics that is largely restricted to montane tropical rain forests. The genus is clearly distinguished by the presence of a two-sided apical cell, gynoecia surrounded by a single, encircling bract, and a unique combination of ornamentation types on the inner and outer walls of the sporophyte capsule. Saccate-lobuled leaves with a valve-controlled aperture are a distinctive feature of many species. Two of the 11 species occur in the neotropics: *P. paradoxa* and *P. heterophylla*. Although few in number, the neotropical species are key to an understanding of the evolutionary history and biogeography of the genus. Unfortunately, both are incompletely known. *Pleurozia paradoxa*, which occurs in the Andes and the Guyana Highlands, is known only from male plants. Until the morphology of gynoecial plants is known, the taxonomic position of this species (as a separate genus or the basal most subgenus) remains open to question. *Pleurozia paradoxa* is most common in the northern Andes and the Guyana Highlands, but some collections are known from southern Chile. The current distribution could have been attained by migration south from North America during the Pliocene via Central America, or north from southern South America. The directionality of this migration would indicate whether *Pleurozia* originated in Gondwanaland or Laurasia. *Pleurozia heterophylla*, known only from a few rather sparse collections on Mt. Roraima, is placed in subgenus *Diversifolia*. This species is very closely related to, and possibly conspecific with an Asian species, *P. subinflata*. Because there are so few collections of this species available, it is impossible to determine if it is a distinct species or if its current distribution is relictual or the result of long distance dispersal. A better understanding of the morphology and distribution of *P. heterophylla* would provide some clues about the timing of speciation in *Pleurozia*.

Barbara M. Thiers, New York Botanical Garden, Bronx, NY 10458-5126, USA. Fax 718-220-6504, Email: bthiers@nybg.org.

LECTURE

Moss Communities in the Páramo and Upper Montane Forest of Cerro de la Muerte, Costa Rica

L. Trujillo

The distribution and abundance of terrestrial mosses were studied in six different types of vegetation, in association with different life-forms of vascular plants, in páramo and at the upper edge of montane forest on Cerro de la Muerte and Cerro Asunción in the Cordillera de Talamanca, Costa Rica, at 2500–3325 m elevation. The vegetation types included: 1) shrubby vegetation dominated by Comarostaphylis, Diplostephium, and Buddleja; 2) thickets of the dwarf bamboo Chusquea which occurs together with low-growing cushion-plants; 3) open areas dominated by bunchgrasses Agrostis, Cortaderia, and Festuca;   4) tall shrubby vegetation dominated by Vaccinium and Myrsine; 5) the upper edge of Quercus forest with low shrubs of Arcytophyllym and Pernettya; 6) open areas of bare soil or rock, affected by frequent fires and other human-caused disturbance, where vascular plants were absent but mosses were very abundant. The moss communities found in association with each type of vascular plant vegetation were analyzed within two transects of 100 × 5 m. Frequency and cover were determined for each moss species within the transects. Biomass was estimated by removing and weighing all mosses in a 1 m² square within each transect. The distribution of mosses in the different vegetation types was analyzed by a similarity index. Forty species of mosses were recorded at the site. Some species, such as Thuidium delicatulum and Polytrichum juniperinum were abundant and widely distributed in all vegetation types. Others, such as Rhacomitrium crispulum, Prionodon densus, and Rhacocarpus purpurascens, were restricted to the bare soil areas of the upper páramo. The tall shrub zone had the highest number of moss species (11) restricted to that habitat. Overall, the most abundant moss species were Porotrichum cobanense, Symplepharis vaginata, and Sphagnum meridense, the latter in association with Breutelia subarcuata. The highest moss biomass occurred in association with Chusqea thickets and open bunchgrass areas. Biogeographically, the moss flora of the Cerro de la Muerte site shows strong affinities with the bryophyte floras of montane Mexico as well as South America.

Luisa Trujillo C., Herbario Nacional del Ecuador, Casilla 17-12-867, Quito, Ecuador. Telephone and Fax 593-2-441592.

LECTURE

Composition and Structure of a Dry Cloud Forest of North-western Peru

N. Valencia

Eight 20 × 20 m plots were studied in a dry cloud forest on the north-western slopes of the Peruvian Andes (departamento Piura), at 2890–2950 m elevation. All trees and shrubs ≥ 2.5 cm dbh were measured and identified. I found 32 families, 40 genera and 52 species. The families represented by the greatest number of species were Compositae and Melastomataceae. The mean density was 2965 stems per hectare, with the main contribution by *Symplocos* sp. (16%), *Myrcine* sp. (13%) and *Hedyosmum scabrum* (11%). The basal area was 79.9 m^2 per hectare with the highest contribution by *Weinmannia reticulata* (22%), *Podocarpus oleifolius* (11%) and *Weinmannia* sp. (11%). The canopy was 13 m tall with emergent trees reaching 22 m.

Niels Valencia, Museo de Historia Natural, Apartado 14-0434, Lima 14, Peru.

LECTURE

Composition and Structure of a Montane Forest on Eastern Ecuador

R. Valencia

In a one-hectare study plot, located at an elevation of 2000 meters at Baeza, all trees with dbh ≥ 5 cm were analyzed according to standard methods. The plot contained 1622 individuals with a total basal area of 39 m² and approximately 50 species. Fourteen species had multiple stemmed trees. The palms were the most conspicuous trees, especially *Geonoma weberbaueri* represented by 800 individuals in the plot. The Lauraceae is represented by several conspicuous trees that sometimes reach 30 meters. The results show a forest with species belonging to lowland forests (*i.e.,* *Cedrela odorata*) mixed with Andean species (*i.e.,* *Brunellia* sp.) and species specific to this altitude (*i.e.,* *Juglans neotropica*) coexisting in this forest. The physiognomy of the forest resembles that of higher Andean forest in the great number of multi-stemmed trees, a feature which is uncommon in lowland forest.

Renato Valencia, Pontificia Universidad Católica del Ecuador, Departamento de Biología, Herbario QCA, Apartado 17-01-2184, Quito, Ecuador. Telephone 593-2-529-250 ext. 279, Fax 593-2-567-117.

LECTURE

Global Change, Biodiversity and Conservation of Neotropical Montane Forests

T. van der Hammen

The purpose of this talk and publication is to explore the relations of the global change and biodiversity and their meaning for conservation policies, in the northern Andean neotropical montane forest zones. Global change of the past first includes tectonically induced changes, resulting in the creation of the northern neotropical montane environments and in the creation of pathways form the northern and southern temperate areas of the Earth, towards the (warm) temperate, cool and cold montane zones of the tropical Andes. Secondly, it includes primary climatic changes (general cooling during the Upper Tertiary and the changes of temperature and rainfall of the Pleistocene glacials and interglacials), affecting heavily the created montane environments and ecosystems, eventually leading to the appearance or disappearance, respectively, introduction of (new) taxa and/or re-ordering of community composition. Certain parts of the northern Andes were already hill-ridges (probably not higher than 500–1000 m) in the early and middle Tertiary, as a result of successive phases of tectonic compression and upheaval. In the Miocene, hills up to 1000 m and maybe up to 2000 m might have been present, and the three Cordilleras and the inter-Andean valleys between them, were already differentiated. Sub Andean (lower montane) forests may then have been developed locally already. The mayor upheaval of the Andes, with highest elevations between 5000–6000 m, took place in the Pliocene, between 5 and 3 million years before present. Extensive lower and higher montane forest zones were then created at the same time that the Panama land bridge was formed. Connections with the northern and southern warm to cool temperate areas (and austral-antarctic) became then optimal, leading to an increased bio-geographical diversity of the montane forests (including elements that were part of the Tertiary-Laurasian flora, nowadays amphi-pacific in distribution). *Hedyosmum* entered the neotropical montane area some four million years ago, *Myrica* some three million years ago, *Alnus* one million years ago, and *Quercus* only some 200,000 years ago. The present biogeographic diversity of the Andean forests seems to increase with altitude, being highest in the upper montane forest zone. The alpha diversity maximum for mosses is around 2000 m, for liverworts at *ca.* 3000 m, and earthworms have their maximum around 3000 m. The interval between 2000 and 3000 m, including the boundary of upper and lower montane forests, seems therefore to have the highest biodiversity. Alpha biodiversity is correlated with present-day rainfall and altitude (temperature), and relative humidity. Beta-diversity may be considerable, when comparing, for example, the East and West slopes of the Central Cordillera, having not a single species of earthworms in common. Relative geographic isolation, partly combined with climatic differences, should play an important role in these cases, as

also the tectonic history.   Conservation of the alpha, beta and biogeographical diversity of the montane neotropical forests (flora, vegetation,
fauna) in relation to present-day global change (climatic warming and
related changes of humidity, and deforestation), requires the taking into
account of the considerable beta-diversity (a result of tectonic and climatic
history), and the climatic gradients (temperature, rainfall and humidity).
Reservations would have to be selected for every biogeographically different
area, and preferably should be areas continuous from above the tree line to
the lowlands, or at least with connecting corridors of natural vegetation,
along temperature, rainfall and humidity gradients.

*Thomas van der Hammen, Tropenbos-Colombia, c/o Corporación
Araracuara, Calle 20 N 5-44, Apartado aéreo 036062, Santafé de Bogotá,
Colombia. Telephone 57-1-281-1616 and 57-1-283-6755, Fax 57-1-286-2418.*

LECTURE

Environmental and Ecological Effects of the Coca Conversion Industry

D. R. Van Schoik and S. H. Schulberg

There has been relatively little research done to assess the environmental effects of coca agriculture due to the illicit nature of the investigated activity and the potential danger associated with such research. Coca leaf is processed into cocaine paste and base in many small laboratories, primarily in Bolivia and Peru. The precursor chemicals used in the conversion process are routinely dumped into streams and rivers. This research sought to quantify the impact of these practices based on a literature review and the results of a preliminary environmental assessment. The study area is in the Chapare, a lush subtropical rain forest region in central Bolivia. The region is considered one of the best coca growing areas in the lower Andes. Up to seven meters of rain swell the rivers, water the forests, and make the Chapare one of the most diverse tropical areas in the world. The environmental and ecological effects described were determined based on analysis of soil samples, measurement of ambient air petroleum hydrocarbon levels, and botanical surveys. The magnitude of these impacts was quantified through comparison of ecological and environmental indices between undisturbed control sites and coca conversion laboratory sites. The results of the field study indicate that deforestation is the most significant impact detected. Each coca conversion laboratory and cultivation plot creates significant localized impacts, resulting in "severely impaired" ecosystem function corresponding to a >75% loss of biological richness, diversity, and abundance. Organic chemical soil contamination is present at all the laboratory sites studied. However, contaminant migration is limited to a radius of approximately four meters from the laboratory site. Coca agriculture is found to reduce soil pH, concentrate aluminium to potentially toxic levels for other crops, and result in massive nutrient export (leaf harvest) which severely depletes the soil. At present, deforestation in the Chapare is not yet severe as major portions of the region remain undisturbed. However, much of the most recent deforestation is occurring in pristine and undeveloped areas of Bolivia as counter-narcotics pressure has the unintended effect of pushing coca growers into peripheral areas. Chronic low level chemical contamination appears to be slowing the natural revegetation process. The release of caustic and toxic chemicals disrupts nutrient recycling and slows revegetation. Coca leaf harvesting robs the ecosystem of vital nutrients by preventing leaf decomposition. This depletion of soil nutrients retards re-establishment of primary forest and slows regrowth of secondary forest in fallow coca fields.

Rick Van Schoik and Seth Schulberg, Southwest Research Associates, Inc., 2006 Palomar Airport Rd., Carlsbad, CA 92008, USA. Telephone 619-431-5640, Fax 691-431-8964.

LECTURE

Biogeography of Neotropical Cloud Forests

G. L. Webster

Neotropical cloud forests extend from 23°N to 25°S latitudes; in the north (Mexico) they are isolated and surrounded by xeric vegetation, whereas in the south (Chile and Argentina) they grade into temperate rain forests. Elevationally, typical, or model, cloud forest is generally found between 1000 and 3000 m, although there is considerable local variation. In terms of climate, cloud forest may be regarded as a subset of montane rain forest in which seasonal variation in precipitation is minimal, and there are no months in which evapotranspiration exceeds rainfall. Total precipitation is generally between 2000 mm and 4000 mm per year, and by definition cloud cover is very nearly continuous. Mean annual temperature decreases with elevation in a continuous gradient from 18–22°C at 1000 m elevation to less than 10°C at 3000 m. With some arbitrariness, a warmer lower montane rain forest may be distinguished floristically from an upper montane rain forest. The lower montane cloud forests, at 1000–2000 m, corresponding to sea-level subtropical rain forests, are characterized by tropical elements such as *Ficus*, *Melastomataceae*, *Annonaceae*, palms, and *Cyclanthaceae*. Upper montane cloud forests are similar in many respects (such as high percentages of epiphytes and ferns), but differ in the presence of distinctly temperate taxa such as *Podocarpus*, *Alnus*, *Drimys*, *Weinmannia*, and Magnoliaceae. The geographic distribution patterns of neotropical cloud forest taxa have been strongly influenced by the Pliocene reconnection between North America and South America, which permitted reciprocal migrations of "Andean" taxa (such as *Drimys*) north and "Cordilleran" taxa (*e.g., Juglans*) south. Pleistocene climatic fluctuations also have strongly affected cloud forests in elevational placements, and have led to marked refugial and archipelagal effects. Endemism in cloud forest taxa is low at the generic level but high at the species level, suggesting recent and rapid speciation.

Grady L. Webster, University of California, Davis, California, USA.

LECTURE

Speciation, Diversity, and Relationships of Theaceae and Bonnetiaceae in the Neotropics

A. L. Weitzman

Theaceae *s. str.* include about 20 genera and perhaps 1000 species centered in the cooler areas of the tropics, especially in Asia, the Pacific, and the Neotropics. In the Neotropics there are six genera, *Cleyera*, *Freziera*, *Gordonia* (formerly *Laplacea*), *Pelliciera*, *Symplococarpon*, and *Ternstroemia*. *Freziera* (56 sp.) and *Ternstroemia* (70 sp.) are the largest genera. *Freziera* is most diverse in the Colombian and Bolivian Andes. *Ternstroemia* is widespread and most diverse in the Venezuelan Guayana, Andes, and C America. Two major subfamilies are recognized, the Camellioideae, represented by *Gordonia* and *Franklinia* in the new world, and the Ternstroemioideae, represented by *Cleyera*, *Freziera*, *Symplococarpon*, and *Ternstroemia* in the Neotropics. *Pelliciera* is not easily placed in either subfamily and its relationships are a subject for future studies. Both subfamilies have their centers of diversity in China. *Gordonia*, *Ternstroemia*, and *Cleyera* have species in Asia. Ternstroemioideae is further divided into two tribes, Ternstroemieae and Freziereae. The Freziereae is represented in the Neotropics by the three remaining genera. *Freziera* is related to the predominantly Asian genus, *Eurya*, especially the Hawai'ian species. The neotropical species of *Cleyera* are clearly related to *Symplococarpon*, while the relationships of the two genera to remaining Freziereae (including the Asian species of *Cleyera)* are less well known. The Bonnetiaceae, which has often been included in Theaceae, has three genera and 33 species centered in the Neotropics. The Bonnetiaceae is more closely related to the Clusiaceae than to the Theaceae. *Ploiarium*, with three species in Southeast Asia, is related to *Archytaea*, with two species of the Guayana region. *Bonnetia*, which is centered in the Venezuelan Guayana, has 28 species. Bonnetiaceae dominate white sand lowland areas and many tepui summits in the Guayana. Both Theaceae and Bonnetiaceae have a few widespread species and many narrow endemics. *Freziera*, for example has several widespread, common species, but 14 of its 56 species are known only from their type localities. Similarly, 14 of the 28 species *Bonnetia* are known from only one tepui. Bonnetiaceae are rare outside the Guayana region, but they dominate shrubby and forested ecosystems there. Many, but not all, of those areas already have some protection. Theaceae rarely dominate tropical montane forests, but because they are often rare, they may be important indicators for regional conservation policies.

Anna L. Weitzman, Department of Botany, NHB-166, Smithsonian Institution, Washington, DC 20560, USA. Telephone 301-299-7762 or 202-357-4808, Fax 202-786-2563, Email: mnhbo064@sivm.si.edu.

LECTURE

The Development of the Colombian Montane Forest as a Consequence of the Late Tertiary Andean uplift

V. M. Wijninga

The vegetational history of the Colombian forests in the Eastern Cordillera is the subject of a combined paleobotanical-palynological study. This study provides the opportunity to investigate the development of the vegetation and the evolutionary adaptation of the flora on a more detailed scale. During the Pliocene (6–2.5 Ma) the area of the high plain of Bogotá (Eastern Cordillera; 2600 m) was affected by the final major uplift as a result of tectonic activity. The area became uplifted from below 500 m to its present altitude of 2600 m. The studied outcrops (present altitude between 2450–2800 m) on the high plain provide a biostratigraphical frame work of in total seven biozones based upon the successive immigration of arboreal pollen taxa. Biozones I–IV represent the actual uplift phase. The outcrops represent successive phases in this uplift and reveal the corresponding paleovegetation. Sediments of Early Pliocene age were deposited under tropical lowland conditions (biozone I; age 5.3 Ma). The paleovegetation is dominated by both microfossils and macrofossils taxa with tropical lowland affinities. The paleoaltitude of these sediments was below 500 m alt. Biozone II represents the first phase in the actual uplift. In the lower part of this biozone the paleovegetation is represented by no modern analog vegetation. The regional vegetation is dominated by microfossils with sub-Andean and Andean affinities. Macrofossils of Humiriaceae and Chrysobalanaceae however, suggest a tropical lowland environment not above 1100 m. The upper part of this biozone (age 3.7 Ma) revealed a paleovegetation dominated by microfossils with sub-Andean affinities. The macrofossils are still under study. Suggested paleoaltitude: 1500 m. Biozone III represents the Late Pliocene. The vegetation is dominated by microfossils with Andean and sub-Andean affinities. The recorded macrofossils have similar affinities. Comparable pollen associations can be found nowadays near the transition between the Andean and sub-Andean vegetation belts, presently located at 2300 m. Macrofossils of *Hippocratea/Pristimera* suggest a paleoenvironment not above 1800 m. In biozone IV (age 2.7 Ma) the paleovegetation is predominantly Andean. At that time the major uplift has ceased and the high plain has reached its present-day altitude of 2600 m.

Vincent M. Wijninga, Hugo de Vries Laboratory, University of Amsterdam, Kruislaan 318, 1098 SM Amsterdam, The Netherlands. Telephone 31-205257672, Fax 31-205257715.

LECTURE

Biogeography of Neotropical Montane Forests: A Reappraisal of Paradigms

K. Young

Biogeography attempts to explain the changing distributions of plants and animals. Recently many researchers have chosen to limit biogeographical explanations to processes affecting taxonomic entities, such as species or genera. I suggest that it is more useful to make biogeography's domain the study of distributions of all levels of biological organisation, from genes to global processes. I use this holistic perspective to examine three paradigms in relation to neotropical montane forests and their biota. These three paradigms concern, respectively, ecological, evolutionary, and anthropogenic processes. Spatial changes in ecological processes are perhaps most usefully studied along environmental gradients such as elevation or as related to regimes of environmental factors or disturbances, such as the persistence of fog or slope instability. Other important spatial concerns for montane forests include habitat fragmentation and connectivity. Because distributions of species in mountains are typically narrow and linear, gene flow is often intermittent. In general, it appears that species interactions here are less coevolved and less predictable than in lowland tropical forests. Evolutionary processes probably have been particularly affected in terms of rates and directions by shifts in elevational zones due to climatic change and orography. Speciation would appear to be at a maximum here, although extinction is probably also quite high compared to the lowland tropics. These ecological and evolutionary patterns and processes make the montane forest biota particularly difficult to protect from human influences, which exasperate the natural fragmentation and discontinuities of montane forests, decrease connectivity among habitat islands, and cause selective extinctions due to overexploitation. Some highland areas have been europeanized by the introduction of exotic species and the development of anthropogenic landscapes. Collectively these three paradigms form a coherent research agenda for the scientific community interested in theoretical and practical insights about the past, present, and future distribution of montane forests in the Neotropics.

Kenneth R. Young, Department of Geography, University of Maryland Baltimore County, Baltimore, MD 21228, USA. Telephone 410-455-3078, Fax 410-455-1056.

Taxonomy, Geographic Distribution and Status of the Genus Cinchona in Peru

P. A. Zevallos Pollito

This study was carried out throughout Peru but with strongest emphasis in northern Peru and in such areas for which reports on the occurrence of the genus *Cinchona* existed. The purpose was to study the taxonomy, geographic distribution and status of the species of *Cinchona*. The existing literature was studied and specimens from the major Peruvian herbaria were examined. Using the methodology proposed by the International Union for Conservation of Nature, fieldwork was carried out in order to collect new material and verify the situation of the species' populations in the field. In total 17 species, previously reported for Peru (Macbride, 1936; Hodge, 1947) were studied; of these, five were found in northern Peru (*C. glandulifera, C. humboldtiana, C. Micrantha, C. officinalis, C. pubescens*).

Percy A. Zevallos Pollito, Facultad de Ciencias Forestales, Universidad Nacional Agraria La Molina, Lima, Peru.

REPORTS FROM THE BOTANICAL INSTITUTE,
UNIVERSITY OF AARHUS
1. **B. Riemann:** Studies on the Biomass of the Phytoplankton. 1976. Out of print.
2. **B. Løjtnant & E.** **Worsøe:** Foreløbig status over den danske flora. 1977. Out of print.
3. **A. Jensen & C. Helweg Ovesen (Eds.):** Drift og pleje af våde områder i de nordiske lande. 1977. 190 p. Out of print.
4. **B. Øllgaard & H. Balslev:** Report on the 3rd Danish Botanical Expedition to Ecuador. 1979. 141 p. Out of print.
5. **J. Brandbyge & E. Azanza:**   Report on the 5th and 7th Danish-Ecuadorean Botanical Expeditions. 1982. 138 p.
6. **J. Jaramillo-A. & F. Coello-H.:** Reporte del Trabajo de Campo, Ecuador 1977—1981. 1982. 94 p.
7. **K. Andreasen, M. Søndergaard & H.-H. Schierup:** En karakteristik af forureningstilstanden i Søbygård Sø — samt en undersøgelse af forskellige restaureringsmetoders anvendelighed til en begrænsning af den interne belastning. 1984. 164 p.
8. **K. Henriksen (Ed.):** 12th Nordic Symposium on Sediments. 1984. 124 p.
9. **L. B. Holm-Nielsen, B. Øllgaard & U. Molau (Eds.):** Scandinavian Botanical Research in Ecuador. 1984. 83 p.
10. **K. Larsen & P. J. Maudsley (Eds.):** Proceedings. First International Conference. European-Mediterranean Division of the international Association of Botanic Gardens. Nancy 1984. 1985. 90 p.
11. **E. Bravo-Velasquez & H. Balslev:** Dinámica y adaptaciones de las plantas vasculares de dos ciénagas tropicales en Ecuador. 1985. 50 p.
12. **P. Mena & H. Balslev:** Comparación entre la Vegetación de los Páramos y el Cinturón Afroalpino. 1986. 54 p.
13. **J. Brandbyge & L. B. Holm-Nielsen:** Reforestation of the High Andes with Local Species. 1986. 106 p.
14. **P. Frost-Olsen & L. B. Holm-Nielsen:** A Brief Introduction to the AAU - Flora of Ecuador Information System. 1986. 39 p.
15. **B. Øllgaard & U. Molau (Eds.):** Current Scandinavian Botanical Research in Ecuador. 1986. 86 p.
16. **J. E. Lawesson, H. Adsersen & P. Bentley:** An Updated and Annotated Check List of the Vascular Plants of the Galapagos Islands. 1987. 74 p.
17. **K. Larsen:** Botany in Aarhus 1963 - 1988. 1988. 92 p.

AAU REPORTS:
18. Tropical Forests: Botanical Dynamics, Speciation, and Diversity. Abstracts of the AAU 25th Anniversary Symposium. Edited by **F. Skov & A. Barfod.** 1988. 46 pp.
19. Sahel Workshop 1989. University of Aarhus. Edited by **K. Tybirk, J. E. Lawesson & I. Nielsen.** 1989.
20. Sinopsis de las Palmeras de Bolivia. By **H. Balslev & M. Moraes.** 1989. 107 pp.
21. Nordiske Brombær (Rubus sect. Rubus, sect. Corylifolii og sect. sect. Caesii). By **A. Pedersen & J. C. Schou.** 1989. 216 pp.
22. Estudios Botánicos en la "Reserva ENDESA" Pichincha - Ecuador. Editado por **P. M. Jørgensen & C. Ulloa U.** 1989. 138 pp.

23. Ecuadorean Palms for Agroforestry. By **H. Borgtoft Pedersen & H. Balslev.** 1990. 120 pp
24. Flowering Plants of Amazonian Ecuador - a checklist. By **S. S. Renner, H. Balslev & L. B. Holm-Nielsen,** 1990. 220 pp.
25. Nordic Botanical Research in Andes and Western Amazonia. Edited by **S. Lægaard & F. Borchsenius,** 1990. 88 pp.
26. HyperTaxonomy - a computer tool for revisional work. By **F. Skov,** 1990. 75 pp.
27. Regeneration of Woody Legumes in Sahel. By **K. Tybirk,** 1991. 81 pp.
28. Régénération des Légumineuses ligneuses du Sahel. By **K. Tybirk,** 1991. 86 pp.
29. Sustainable Development in Sahel. Edited by **A. M. Lykke, K. Tybirk & A. Jørgensen,** 1992. 132 pp.
30. Arboles y Arbustos de los Andes del Ecuador. By **C. Ulloa Ulloa & P. M. Jørgensen,** 1992. 264 pp.
31. Neotropical Montane Forests. Biodiversity and Conservation. Abstracts from a Symposium held at The New York Botanical Garden, June 21–26, 1993. Edited by **Henrik Balslev,** 1993, 110 pp.

ORDER FORM

To Aarhus University Press
Aarhus University
DK-8000 Aarhus, DENMARK

I would like to order the following issues of AAU REPORTS
at 78 DKr (13 USD) each:

_____

Please send the books to:

Name:

Street:

Town:

Country:

Do not send payment with order; we will bill you later